Making Crime Pay

STUDIES IN CRIME AND PUBLIC POLICY
Michael Tonry and Norval Morris, *General Editors*

Police for the Future
David H. Bayley

Incapacitation: Penal Confinement and the Restraint of Crime
Franklin E. Zimring and Gordon Hawkins

The American Street Gang: Its Nature, Prevalence, and Control
Malcolm W. Klein

Sentencing Matters
Michael Tonry

The Habits of Legality: Criminal Justice and the Rule of Law
Francis A. Allen

Chinatown Gangs: Extortion, Enterprise, and Ethnicity
Ko-lin Chin

Responding to Troubled Youth
Cheryl L. Maxson and Malcolm W. Klein

Making Crime Pay: Law and Order in Contemporary American Politics
Katherine Beckett

Making Crime Pay

Law and Order in Contemporary American Politics

KATHERINE BECKETT

New York Oxford
Oxford University Press
1997

Oxford University Press

Oxford New York
Athens Auckland Bangkok Bogota Bombay Buenos Aires
Calcutta Cape Town Dar es Salaam Delhi Florence Hong Kong
Istanbul Karachi Kuala Lumpur Madras Madrid Melbourne
Mexico City Nairobi Paris Singapore Taipei Tokyo Toronto Warsaw

and associated companies in
Berlin Ibadan

Published by Oxford University Press, Inc.
198 Madison Avenue, New York, NY 10016

Library of Congress Cataloging-in-Publication Data
Beckett, Katherine, 1964–
Making Crime Pay: Law and Order in Contemporary American Politics / Katherine Beckett.
p. cm. — (Studies in crime and public policy)
Includes bibliographical references and index.
ISBN 0-19-511289-X
1. Crime—Political aspects—United States. 2. Criminal justice, Administration of—Political aspects—United States. 3. Crime prevention—Political aspects—United States 4. Narcotics, Control of—Political aspects—United States. 5. United States—Politics and government—1989– I. Title. II. Series.
HV6791.B42 1997
364.973—dc20 96-31521

9 8 7 6 5 4 3 2 1

Printed in the United States of America
on acid-free paper

Acknowledgments

I am privileged to have benefited from the insight and support of many mentors, colleagues, and friends throughout the course of this project. The analysis presented in chapter 2 would not have been possible without the assistance of Bruce Western; his knowledge, generosity, and commitment to collegiality are inspirational to me. Ivan Szelenyi's interest in power and the production of knowledge shaped the development of this book's epistemological framework, and his willingness to invest time and energy in it was invaluable. Craig Reinarman's insight regarding the political and cultural aspects of the drug issue and his insistence that I conceive of my dissertation as a book-in-progress were tremendously helpful. I was also fortunate to have access to Bill Roy's expertise in political and historical sociology; his enthusiasm and empathy were a crucial source of support throughout my graduate career. Frank Gilliam made an important contribution by helping to clarify the ways in which contemporary electoral dynamics and the racialization of American politics inform the politicization of crime-related issues.

Theodore Sasson has been a wonderful source of ideas and constructive criticism; his willingness to wade through the various incarnations of this project and his unflagging enthusiasm for it are much appreciated. Steve Sherwood provided many opportunities for me to clarify and develop my ideas; even when disagreeing, he was encouraging.

Joachim Savelsberg identified important conceptual issues and asked thought-provoking questions along the way. Thanks to Joe Nevins for suggesting the title and for supporting this project since its inception. I am also indebted to the editors of this series—Norval Morris and Michael Tonry—for their interest and feedback and to several additional anonymous reviewers of the manuscript.

Steve Herbert has been a continual source of advice, ideas, and criticism; I am fortunate to have him as an intellectual and life partner. Finally, I would like to thank my mother: through her example of compassion and conviction, Joyce Beckett inspired the faith in the possibility of redemption upon which this book rests.

Contents

Making Crime Pay

1

Law and Order in Contemporary American Politics

Crime and punishment sit center stage in the theater of American political discourse. For much of the past three decades, politicians have made crime-related problems central campaign issues and struggled to identify themselves as tougher than their competitors on crime, delinquency, and drug use. Popular concern about these social problems has reached record levels during this period[1] and public opinion polls indicate that members of the public have become more likely to support punitive policies such as the death penalty and "three-strike" sentencing laws.[2] Not surprisingly, these ideological shifts have been accompanied by a dramatic expansion of the criminal justice system. Between 1965 and 1993, crime control expenditures jumped from $4.6 billion to $100 billion (and from .6 to 1.57% of the gross domestic product) and the rate of incarceration in the United States is now the highest in the industrialized world.[3] Minorities have been especially affected by these developments: blacks now comprise over half of all prison inmates in the United States, up from one-third just twenty years ago.[4]

How did we get here? Why have crime-related problems assumed such prominence in recent decades, and what accounts for the insistence that harsher punishments and tougher law enforcement are the best response to these complex social problems? Despite its importance, this question has not been addressed as systematically as one might expect. To the extent that it has been, most analysts have offered a fairly simple explanation: concern about crime and punitive attitudes are widespread because the crime and drug problems have gotten worse. According to this "democracy-at-work" thesis,[5] the increased incidence of criminal behavior has led Americans to demand that their political representatives crack down on criminals; the more frequent use of the death penalty and the adoption of tough three strikes sentencing laws are politicians' responses to this popular sentiment. In sum, this thesis suggests that the current approach to crime control reflects the worsening of the crime problem and the public sentiment to which this trend naturally gives rise.

Although intuitively appealing, this explanation does not withstand closer examination. Proponents of the democracy-at-work thesis typically point to official crime statistics which indicate that the rate of crime increased throughout the 1960s and 1970s. But as we will see in the following chapter, levels of public concern about crime and drug use are not consistently associated with the reported incidence of these social problems. Furthermore, the assumption that anxiety about crime drives support for punitive anticrime policies is problematic. In fact, those who are less afraid of crime typically express the highest levels of support for the "get-tough" approach while those who are more fearful are often less punitive. Rural white men, for example, feel relatively safe but are quite staunch supporters of law and order policies. Conversely, women and blacks are, in general, more concerned about their potential victimization but less supportive of tough crime control measures.[6] The relationship between perceptions of the crime problem and attitudes toward punishment is thus more complicated than the democracy-at-work thesis implies.

Public support for punitive anticrime policies is also more fluid and ambivalent than is commonly supposed. Enthusiasm for the death penalty, for example, is historically variable, weakens considerably in the presence of alternatives, and coexists with widespread support for rehabilitative ideals. When given a choice, most Americans still believe that spending money on educational and job training programs is a more effective crime-fighting measure than building prisons.[7] While the punitive tone of the law and order discourse clearly resonates with salient sentiments in American political culture, popular beliefs about crime and punishment are complex, equivocal, and contradictory, even after decades of political initiative on these subjects.[8] The notion that the desire for punishment is ubiquitous and unequivocal ignores the complexity of cultural attitudes and the situational and political factors that shape their expression.[9]

In sum, support for punitive anticrime measures has waxed and waned throughout American history, coexists with support for less punitive policies, and is only loosely related to the reported incidence of crime-related problems. By positing a direct connection between the incidence of these problems and public punitiveness, the democracy-at-work thesis assumes what requires explanation: the rise of the conception of crime as the consequence of insufficient punishment and control. This book takes this ideological accomplishment as its object of inquiry and tells a very different story about the ascendance of the get-tough approach to crime.

Culture, Politics, and the Construction of Social Problems

Sociologist Max Weber used the term "vielseitigkeit" to refer to the multiplicity of meanings inherent in the social world, a phenomenon he called the "many-sidedness of reality."[10] Others make a similar point when they stress the promiscuous nature of the "ideological sign": because social objects and issues are "multi-accentual" they can acquire a number of different meanings, each of which may have quite distinct political implications.[11] The "crime problem," for example, may be depicted in a variety of ways: as evidence of the breakdown of law and order, the demise of the family, or socioeconomic inequality and the need for policies that reduce it. While the harm victims of crime suffer is very real, our understanding of the meaning and causes of this harm depends upon the way in which the crime issue is apprehended in political discourse.[12] As David Garland concludes, "[I]t is clear enough that criminal conduct does not determine the kind of penal action that a society adopts. . . . [I]t is not 'crime' or even criminological knowledge about crime which most affects policy decisions, but rather the ways in which 'the crime problem' is officially perceived and the political positions to which these perceptions give rise."[13]

Crime-related issues, then, are socially and politically constructed; they acquire their meaning through interpretive, representational, and political processes. Social actors—sometimes called "claimsmakers"[14] —struggle to gain acceptance for preferred ways of framing these issues and vie for limited access to public venues in order to promote them.[15] In these battles over the signification of crime-related problems, claimsmakers "deploy mediated symbols and mobilize powerful cultural references."[16] The Bush campaign's manipulation of the "Willie Horton" incident, for example, can be understood as an attempt to invoke the image of "the black rapist" (with all its historical and cultural significance) in order to generate support for law and order policies—and for the candidate who was, presumably, more capable of implementing them.

Such efforts to signify social problems are typically components of larger political battles. Participants in these broader struggles use a variety of rhetorical devices and cultural images to link ostensibly unrelated social issues in ideologically useful ways. Southern politicians and law enforcement officials who called civil rights protestors "thugs" and decried "crime in the streets," for example, were attempting to define protest activities as "criminal" rather than political in nature. Claimsmakers may also define social problems in ways that direct attention away from inconvenient social conditions. Emphasizing the pathology of criminals and the utility of punishment, for example, obscures the role of social inequality in the generation of crime.[17] Political outcomes such as three strikes legislation are thus best understood as a product of symbolic struggles in which actors disseminate favored ways of framing social problems and compete to have these versions of reality accepted as truth.

These competing "issue frames" are created, mobilized, and institutionalized (or not) under particular historical and political circumstances, and as the Willie Horton incident suggests, officials often play an important role in these campaigns. Elite claimsmaking activities do not merely express popular sentiment but also seek to shape and transform it in accordance with particular visions of state and society.[18] The involvement of officials in these campaigns may be quite consequential: elites often enjoy greater access to public venues, and their proclamations are typically accorded a great deal of authority. President George Bush's (nationally televised) contention that drug abuse constituted "our nation's most serious domestic problem," for example, certainly carried more weight and had greater consequences than would the same statement made by a community activist seeking increased treatment funds. An account of why some representations become institutionalized while others do not thus requires that the analyst move into the realm of power.

Claimsmakers' ability to gain access to the mass media is a particularly important dimension of these power relations because it is through the mass media that issue frames are reproduced and disseminated. While nonelite claimsmakers are sometimes able to influence media coverage,[19] the mutual interdependence of the state and the mass media means that officials are uniquely privileged in the contest to signify social problems. This interdependence is expressed in and reinforced by media practices that lead journalists to rely on political elites for much of their information. The state, in turn, has developed and deployed an elaborate set of institutions aimed at "news management." Officials thus enter contests over social issues with a relatively high degree of access to the mass media and endeavor to maintain and enlarge this advantage vis-à-vis others (some of whom may also be advantaged in this respect).

But access to the mass media does not guarantee the success of claimsmaking enterprises. The capacity of elites to mobilize public opinion depends upon their ability to select symbols and rhetoric that will resonate with deep-seated "myths"[20] and make sense of lived experience. While popular sentiment is somewhat malleable, members of the public are not receptive to every claim and elites are therefore somewhat constrained in their efforts to mobilize opinion. On the other hand, these constraints are far from determinant: "culture" is composed of a variety of often contradictory themes, experiences, and sentiments, and a number of different issue frames may enjoy some cultural resonance at a given historical moment. It is clear, for example, that the discourses of retribution and rehabilitation both enjoy a high degree of support in contemporary American political culture.

The likelihood that competing issue frames will resonate with popular sentiment does not depend upon "expert" opinion, much to the chagrin of some criminologists. Although research may tell us something about the validity of the relationships posited in different crime "frames," this more technical discourse rarely influences the highly symbolic sphere of political rhetoric. Instead, the viability of alternative issue frames rests primarily on the extent to which they help to make sense of people's experience in ways that are compatible with popular wisdom and salient cultural themes.[21] Crime discourse that attributes the criminal behavior of the "underclass" to the expansion of welfare programs is one way of acknowledging the "commonsense" connection between poverty and street crime and simultaneously provides working persons with an explanation for their increasing tax burden. The ability of this discourse to make sense of these "realities" and to identify a target for the anger they induce—rather than the robustness of the regression coefficients designed to measure the strength of the relationships posited—is crucial to the success of this discursive construction.

In sum, sociohistorical context, public discourse, and popular sentiment are related in complex ways. The fact that members of the public tend to express concern about crime-related issues when officials accord them greater attention does not mean that political elites have an unlimited capacity to shape public opinion. Furthermore, it is clear that punitive anticrime rhetoric does resonate with important themes and sentiments in American political culture and provides some with a compelling explanation for pressing social and personal ills. It remains true, however, that political elites have played a leading role in calling attention to crime-related problems, in defining these problems as the consequence of insufficient punishment and control, and in generating popular support for punitive anticrime policies. This book analyzes the origins and nature of this discursive campaign and its consequences for state policy.

The Changing Nature of Public Discourse on Crime

Official perceptions of "the crime problem" have changed dramatically in recent years. For much of the twentieth century, a philosophy and style of reasoning called "penological modernism" served as the foundation of both criminal justice and social welfare practices. According to this philosophy, deviant behavior is at least partially caused (rather than freely chosen). Progressive reformers therefore identified rehabilitation—operationally defined as the use of "individualized, corrective measures adapted to the specific case or the particular problem"—as the appropriate response to deviant behavior.[22] While the goal of rehabilitating offenders often conflicted with competing objectives (especially the hope that punishment would deter individuals from breaking the law), it nonetheless served as the primary rationale for Western crime control policy for much of the twentieth century.[23] Since the 1930's, the modernist, rehabilitative project emphasized environmental theories of crime and therefore provided an alternative to both biological and classical ("free will") explanations of criminal behavior.[24]

The goals and suppositions of this approach are now seen as suspect by many. Where the disappointing results of rehabilitative programs were once regarded as a challenge, the sense that "nothing works" has become widespread and the presumption that criminal behavior has causes that may be identified and remedied by experts has been called into question.[25] Despite the complexity of political discourse on crime, it appears that two main alternative discourses have filled the vacuum created by the demise of the rehabilitative ideology. Among politicians and other officials, policies that promise to enhance deterrence, retribution, and public safety (mainly through incapacitation) are a top priority. These tough responses to the crime problem are predicated upon various (and sometimes contradictory) explanations of criminal behavior: the neoclassical vision of criminals as rational and freely choosing agents, currently undeterred as a result of "undue lenience"; cultural theories that highlight the moral depravity of those who commit crimes (and sometimes the role of "permissive" welfare programs in generating it); and, increasingly, the notion that most criminals are intrinsically—perhaps biologically—"prone to evil" and are therefore beyond redemption. Despite their differences, these explanations of crime similarly imply that expanding the scope of criminal law and increasing the severity of its penalties are the most appropriate responses to the crime problem.

A second crime discourse permeates the writings of criminal justice administrators, penologists, and other practitioners. These experts are largely uninterested in the symbolic dimensions of punishment and focus instead on the need to devise more efficient means of controlling potentially troublesome individuals. Increasingly absent from these discussions is the idea that the crime problem can be "solved" or that the causes of criminal behavior may be identified and remedied.[26] This "administrative" or "managerial" criminology—sometimes called the "new penology"[27]—is technocratic, behaviorist, and "realistic" in tone and is primarily oriented toward devising new and better techniques for managing the crime problem.

In both the politicians' get-tough rhetoric and administrators' managerial criminology, then, the emphasis has shifted from a concern with rehabilitating and reintegrating offenders to the capacity of the law and the social control system to structure the choices and conduct of individuals. This diminution of rehabilitative zeal—what Garland calls "therapeutic nihilism"[28]—is indicative of the more pessimistic mood that characterizes contemporary penology. Accounts of this shift often highlight the role of progressives in unintentionally precipitating the adoption of more retributive and punitive anticrime policies.[29] While liberal and radical critiques of the rehabilitative project developed in the 1970s were undoubtedly influential, the conservative campaign for "law and order" has been more relevant to the ideological and policy shift to the right on crime-related issues. For as Garland suggests, the questioning of the rehabilitative ideal within criminology coincided with "a powerful shift in the political orientation of several Western governments, with the result that penal organizations have been more vulnerable to external political pressures than they might otherwise have been. Indeed, if one were writing a history of penality's present, it is probably here that one would begin."[30]

Crime, Drugs, and the Reconstruction of the State

Since the 1960s, conservatives have paid an unprecedented amount of attention to the problem of "street crime," ridiculed the notion that criminal behavior has socioeconomic causes, and promoted the alternative view that crime is the consequence of "insufficient curbs on the appetites or impulses that naturally impel individuals towards criminal activities."[31] This attempt to reconstruct popular conceptions of the crime problem was, in turn, a component of a much larger political contest: the effort to replace social welfare with social control as the principle of state policy.[32] As the civil rights, welfare rights, and student movements pressured the state to assume greater responsibility for the reduction of social inequalities, conservative politicians attempted to popularize an alternative vision of government—one that diminishes its duty to provide for the social welfare but enlarges its capacity and obligation to maintain social control.[33] In what follows, I show that the crime issue has been a crucial resource for those advocating this reconstruction of social policy.[34] The conservative view that the causes of crime lie in the human "propensity to evil," rests on a pessimistic vision of human nature, one that clearly calls for the expansion of the social control apparatus. Similarly, the notion that the "culture of welfare" causes crime and other behavioral "pathologies" such as addiction, illegitimacy, and delinquency implies the need to scale back the welfare state. Crime-related problems—with all their racial connotations and emotional qualities—have thus been central to the construction of a threatening and undeserving underclass, the emergence of which has done much to legitimate this reconstruction of the state's role and responsibilities.[35]

The Organization of the Book

My emphasis on the political origins and role of the crime issue is clearly at odds with the idea that crime-related attitudes and policies are primarily driven by the incidence of criminal behavior and the public concern that it engenders. The following chapter therefore investigates the relationship between the reported incidence of crime-related problems, levels of concern about and fear of crime, and support for punitive anticrime policies. The results of this analysis suggest that the links between these variables are quite tenuous but that public concern about crime and drugs is strongly associated with prior political initiative on the crime and drug issues. Together, these findings suggest that support for tough anticrime policies is not merely a reaction to the increased incidence of crime and drug use (as indicated by official data) and call attention to the political and ideological processes by which punishment and control have been defined as the primary solutions to crime-related problems.

Chapters 3 and 4 analyze the discursive and political processes through which this was accomplished. The rhetoric of law and order was first mobilized in the late 1950s as southern governors and law enforcement officials attempted to generate and mobilize while opposition to the civil rights movement. As civil rights became a national rather than a regional issue, and as welfare rights activists pressured the state to assume greater responsibility for social welfare, the battle over state policy intensified. At stake was the question of whether the federal government is obligated to assume responsibility for creating a more egalitarian society. Without being explicitly identified as such, competing images of the poor as "deserving" or "undeserving" became central components of this debate. In drawing attention to the problems of street crime, drug addiction, and delinquency, and by depicting these problems as examples of the immorality of the impoverished, conservatives promoted the latter image. The crimes of the poor were thus used as evocative symbols of their undeserving and dangerous nature. The racialized nature of this imagery has been a crucial resource for those attempting to promote this conception and policies that reflect it.

Indeed, race, crime, violence, delinquency, and drug addiction have become defining features of those now referred to as "the underclass."[36] Chapter 4 analyzes the way in which this discourse and the organizational dilemmas associated with the federal government's "war on crime" facilitated the emergence of the antidrug campaign of the 1980s, and pays particular attention to the increased involvement of Democratic party officials in the wars on crime and drugs. This chapter also analyzes the resurgence of anticrime rhetoric in the 1990s and shows that while the identity of the key players in this campaign has changed somewhat, the nature of this rhetoric and the political implications of its ascendance have not.

Chapters 5 and 6 analyze popular support for the wars against crime and drugs and argues that this support (to the extent that it exists) reflects officials' ability to disseminate the discourse of law and order through the mass media as well as its resonance with important cultural themes and sentiments. Chapter 5 uses frame analysis techniques to show that political elites—especially politicians and law enforcement personnel—frequently served as sources in news stories that focused on crime and drugs and that the presence of these sources was strongly associated with the depiction of "issue packages" that identify "liberal permissiveness" and the loss of "respect for authority" as the main causes of crime. While their capacity to shape media representations is not infinite and must be recognized as an achievement (of sorts), officials were quite effective in using the mass media to disseminate images of the crime and drug problems that imply the need for greater punishment and control.

Chapter 6 analyzes popular receptivity to this imagery and suggests that the get-tough discourse does resonate with important sentiments and myths that characterize American political culture. For example, the neoclassical depiction of crime as an individual choice is consonant with the individualism that is so pronounced in American life. Similarly, the argument that welfare programs encourage family breakdown and other "pathologies" resonates with the cultural propensity to attribute social problems to inadequate family life and faulty socialization. Finally, the emotional qualities of the crime issue appear to have enhanced popular support for the law and order campaign.

However, although it is true that the campaign for law and order has been bolstered by these cultural resonances, support for punitive policies is neither unambiguous nor evenly distributed. Survey research indicates that the law and order approach to the crime problem is particularly popular among those who hold racially and socially conservative views. In-depth interviews with such voters reveal that racially charged hostility toward those who "seek something for nothing" is widespread and that this hostility informs support for punitive anticrime policies. Thus, it appears that the "coded" racial subtext of the conservative rhetoric on crime and punishment has not gone unnoticed but has been crucial to its acceptance among these swing voters. The strength of these sentiments has had quite significant political implications: both the Republican and Democratic parties have had their eye on these "Reagan Democrats," among whom punitive crime rhetoric enjoys especially strong support.

Chapter 7 examines the consequences of the federal campaign for law and order and shows how the politicization of the crime issue triggered the expansion and reorientation of the crime control system. In waging the wars on crime and drugs, the federal government has developed a variety of mechanisms that enable it to influence state and local criminal justice policy. The ascendance of the get-tough approach at the national level thus led to the expansion of the entire penal apparatus, which in turn triggered the growth of a politically powerful "penal-industrial complex"[37] that endeavors to perpetuate this expansion. The emergence of the managerial criminology described earlier is also related to the rapid growth of the criminal justice system: this approach is aimed at reducing the fiscal and organizational costs associated with the get-tough approach and promises to do so through the application of cost-effective observational and incapacitative technologies, carefully calibrated according to assessments of risk.[38] Organizational, political, and ideological developments precipitated by the campaign to get-tough on crime and drugs have thus served largely to perpetuate and facilitate that effort. The final chapter reiterates the main outlines of the argument, considers the implications of a state that prioritizes social control over social welfare, and highlights the need for the creation of a more inclusive and pluralistic dialogue regarding crime-related problems.

2

Setting the Public Agenda

Crime and drug use are not naturally or inherently "social control" issues but are constructed as such by social actors; the institutionalization of the get tough approach reflects the ascendance of this interpretation of their causes and solutions. Recognizing the importance of the symbolic dimensions of the crime issue does not imply that crime is not a "real" problem; particularly for the poor and nonwhite, the threat of criminal victimization and the harm associated with drug abuse are all too real. At the same time, the extent to which members of the public express concern about these social problems and, more importantly, become more supportive of punitive anticrime policies is clearly linked to the pervasiveness of imagery and rhetoric that depict these problems as the consequence of excessive lenience.

What came to be known as "the crime issue" emerged on the national political scene during the 1964 presidential campaign and continued to play an important role in national politics through 1972. The reported rate of crime also increased throughout the 1960s; for many, this trend provided more than ample support for the democracy-at-work thesis. By contrast, the war on drugs of the 1980s was waged at a time when the reported incidence of drug use was declining. This chapter investigates this puzzle and summarizes two main bodies of evidence that cast doubt on the conventional interpretation of these events.

The first of these shows that levels of public concern are largely unrelated to the reported incidence of crime and drug use but are strongly associated with the extent to which elites highlight these issues in political discourse. In the next section of the chapter, I summarize a wide body of survey research which suggests that anxiety about crime does not necessarily give rise to punitiveness. Thus, even if concern about or fear of crime were consistently associated with its reported incidence, there is no reason to assume that this would *necessarily* lead members of the public to clamor for the death penalty and stiffer sentencing laws. While the increased incidence of crime-related problems may facilitate their politicization and contribute to growing support for getting-tough, complex cultural processes—in which political elites play a crucial role—clearly shape the formation and expression of popular sentiments regarding crime and punishment.

Evaluating the Democracy-at-Work Thesis

As we have seen, the democracy-at-work thesis holds that the increasing threat of criminal victimization and the anxiety that it engenders explain the adoption of law and order policies. Raymond Michelowski summarizes this argument as follows: "This steady rise in the crime rates . . . generated a growing public fear of crime, a politicization of the crime problem, and eventually political mobilization of this fear of crime turned into demands for more and harsher punishments for lawbreakers. This, in turn, led to a dramatic rise in the absolute numbers of people incarcerated. . . ."[1] Interestingly, this interpretation has been promoted by researchers from across the ideological spectrum. For example, one prominent Marxist criminologist suggested that "[A]s for moral panics about crime in the streets, they were not created by the government. . . . The crime issue was forced on a reluctant Johnson administration by voters exposed to and concerned with crime in their neighborhoods."[2] The well-known conservative James Q. Wilson similarly argued that "public opinion was well ahead of political opinion in calling attention to the rising problem of crime."[3] While these arguments were put forward in an attempt to explain the politicization of crime in the 1960s and 1970s, a similar type of reasoning has been used to explain the war on drugs of the 1980s.[4]

It should be noted that the democracy-at-work thesis is generally presented rather cursorily, as if obvious and not in need of elaboration. To the extent that evidence is cited to support it, proponents of the democracy-at-work thesis generally point to official data sources—especially the Uniform Crime Reports (UCR)—which indicate that the rate of crime increased throughout the 1960s and 1970s.[5] The implicit argument seems to be that these crime data reflect a real increase in the incidence of crime and that as people became aware of this trend (as a result of their own victimization or the victimization of others known to them), they became more concerned about crime and hence more punitive.

Its apparent popularity notwithstanding, the democracy-at-work thesis is in tension with a growing body of literature that stresses the socially constructed nature of social problems such as crime and drug use.[6] Constructionists emphasize that reality is not known directly, but must be comprehended through frames that select, order, and interpret it. These researchers also point out that media personnel and political elites often play an important role in these symbolic processes. A constructionist account of the crime and drug issues therefore anticipates that the public's assessment of the causes and seriousness of social problems will be shaped by public discourse around them. In sum, while the democracy-at-work thesis holds that increases in the incidence of crime and drug use lead members of the public to identify crime or drugs as the nation's most important problems, a constructionist approach emphasizes the impact of political and media discourse on popular attitudes. These alternative hypotheses are evaluated below.

Crime, Drugs, and Public Concern

The following analysis of public concern about crime-related problems is divided into two periods. The first examines public concern about crime during the war on crime (from 1964 to 1974); the second focuses on concern about drug use during the most recent war on drugs (1985–1992).[7] Ordinary least squares (OLS) regression techniques[8] are used to estimate the degree of association between the reported incidence of crime[9] and drug use,[10] on the one hand, and levels of public concern about these social problems[11] on the other. Political initiative[12] on and media coverage[13] of these topics were also analyzed as possible sources of influence on public attitudes. The trend lines for each of these variables are depicted in figures 2.1–2.8.

These figures show that while the reported rates of crime and drug use shifted slowly and gradually, public concern about these problems fluctuated quickly and dramatically.[14] Indeed, in both the crime and

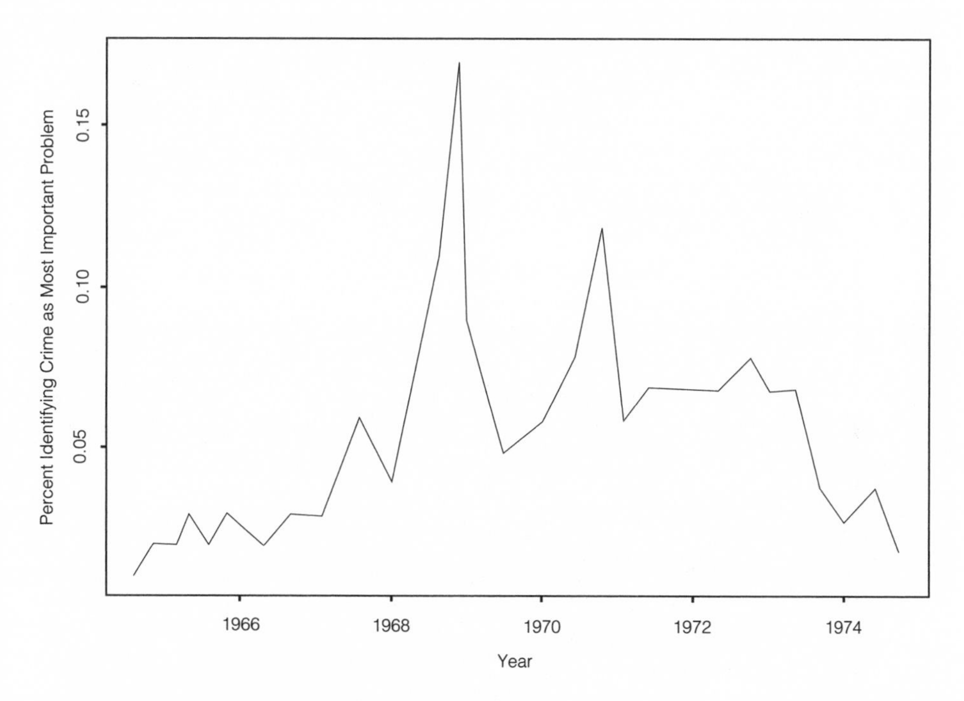

FIGURE 2.1 Public Concern About Crime

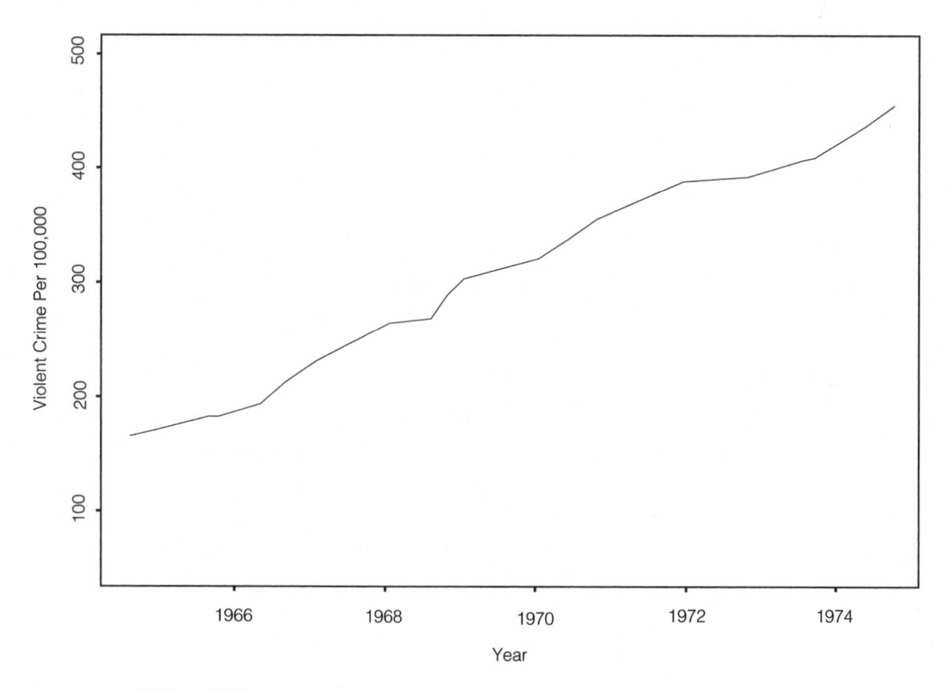

FIGURE 2.2 Crime Rate

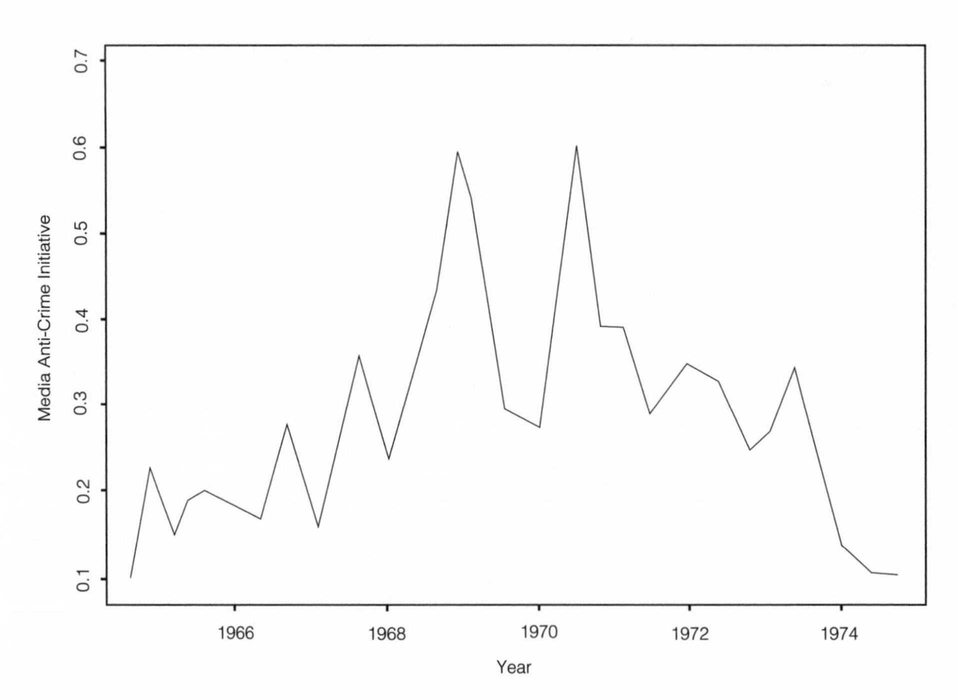

FIGURE 2.3 Media Coverage of the Crime Issue

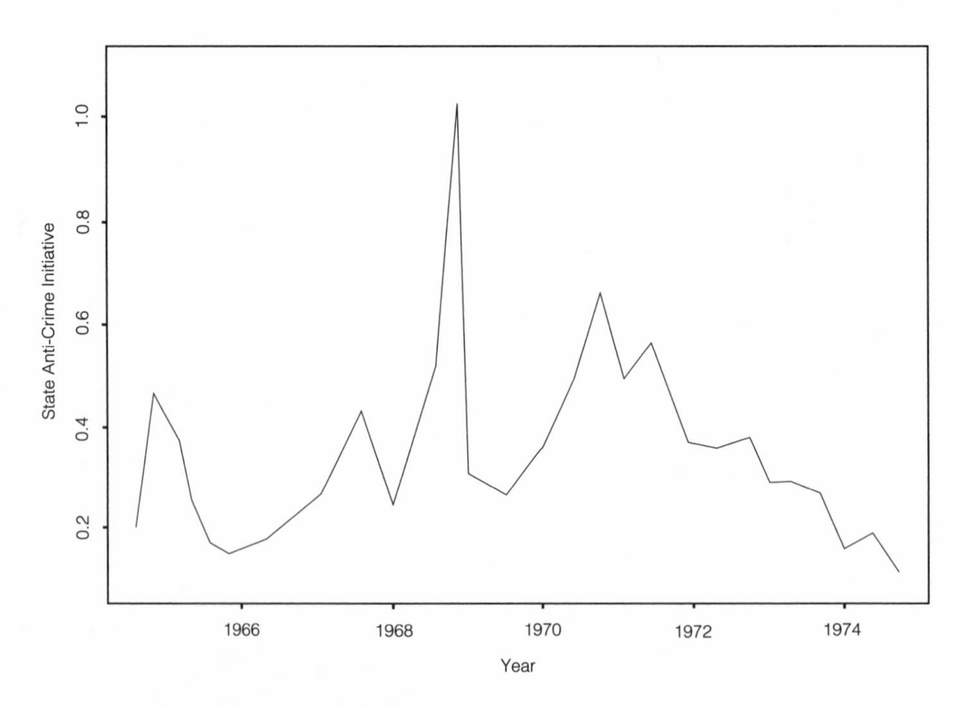

FIGURE 2.4 State Anti-Crime Initiative

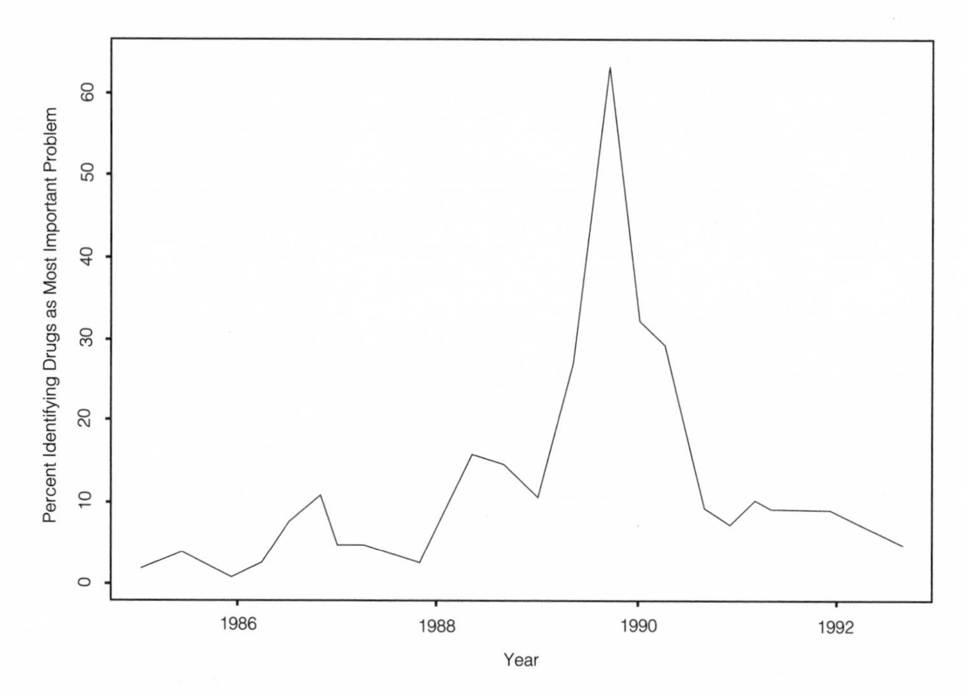

FIGURE 2.5 Public Concern About Drugs

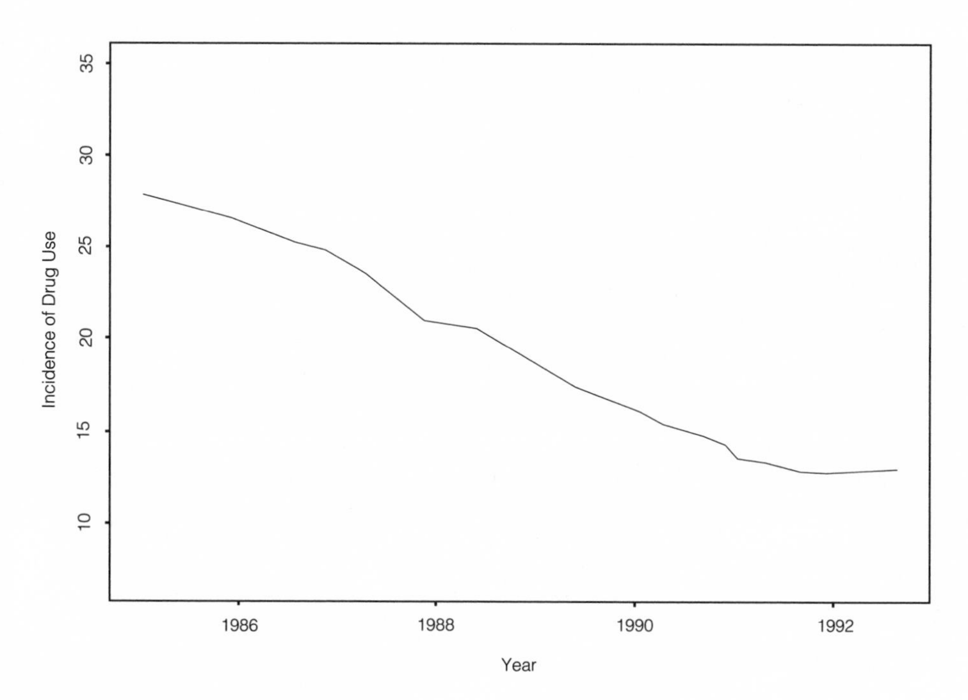

FIGURE 2.6 Incidence of Drug Use

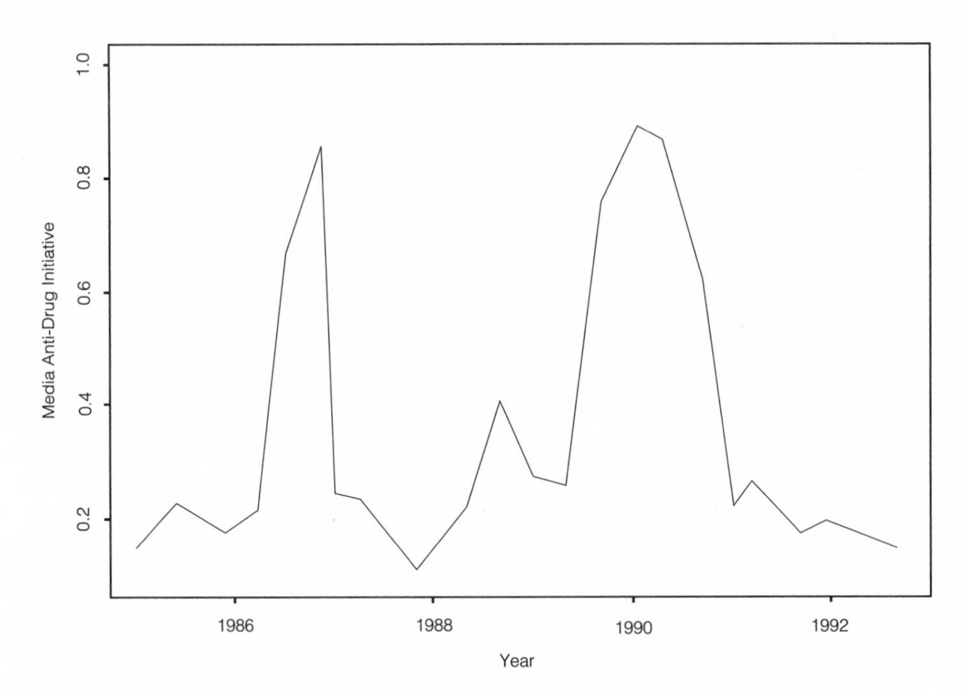

FIGURE 2.7 Media Coverage of the Drug Issue

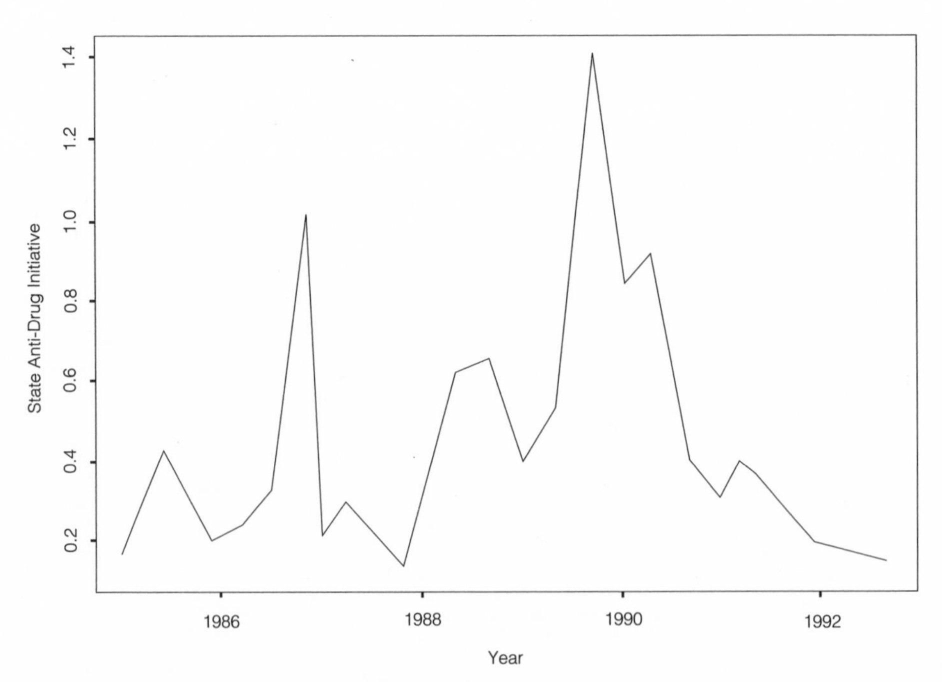

FIGURE 2.8 State Anti-Drug Initiative

drug cases, striking shifts in levels of public concern took place in very short periods of time. For example, the percentage of poll respondents reporting that drugs were the nation's most important problem increased from 15% to 64% between July and September 1989 and dropped to 10% within the following year (see figure 2.5). This sort of fluctuation does not appear to correspond to the reported incidence of crime or drug use, but does seem to be closely related to levels of prior political initiative on the crime and drug issues.

The analysis presented here is aimed at confirming or disconfirming these impressions.[15] The explanatory variables were measured in terms of their average rate in the three- to five-month period preceding each public opinion poll; the (nonlagged) regression results thus indicate the level of association between these variables and immediately subsequent levels of public concern. These regressions were also estimated with a lag of 1 (6–10 months) and 2 (9–12 months) in order to assess their association with delayed shifts in levels of public concern about crime and drugs.

The results of this analysis are presented in tables 2.1 and 2.2.[16] The unstandardized coefficient for each variable is shown, and the standard error (SE) appears beneath it in parentheses. The results in the crime case indicate that both political initiative and media coverage were associated with subsequent levels of public concern about crime (see table 2.1). These relationships are consistent over time: both political initiative and media coverage continue to be significantly and

TABLE 2.1. Correlation of the Crime Rate, Media Coverage, and Political Initiative with Public Concern about Crime, 1964–1974

Explanatory variables	Column 1 Lag = 0 3–5 months	Column 2 Lag = 1 6–10 months	Column 3 Lag = 2 9–15
Crime rate	−.0077 (.011)	−.0067 (.013)	−.005 (.022)
Media initiative	1.2504* (.5547)	1.3103** (.497)	1.2107* (.5372)
Political initiative	1.3711** (.3509)	1.3511** (.3364)	1.2721** (.3409)
Adjusted R^2	.5649	.5866	.5712

*p <. 05

**p <. 01

***p <. 001

TABLE 2.2. Correlation of Rates of Drug Use, Media Coverage, and Political Initiative with Public Concern about Drugs, 1985–1992

Explanatory variables	Column 1 Lag = 0 3–5 months	Column 2 Lag = 1 6–10 months	Column 3 Lag = 2 9–15
Drug use	.0096 (.2178)	.0082 (.1917)	.014 (.2077)
Media initiative	.0594 (.7459)	.0781 (.699)	.0999 (.6781)
Political initiative	1.8393*** (.4551)	1.762*** (.446)	1.1221** (.4997)
Adjusted R^2	.6337	.6291	.6009

*p <. 05

**p <. 01

***p <. 001

positively associated with public concern when an extended time period is analyzed. In contrast, the reported incidence of crime is not associated with the propensity of members of the public to identify crime as the nation's most important problem.[17]

The results presented in table 2.2 indicate that political initiative on the drug issue is positively and significantly associated with subsequent shifts in public concern about drugs.[18] In this case, neither the reported incidence of drug use nor media coverage of the drug issue is associated with levels of public concern about drugs.

One possible explanation for the absence of an association between drug use and public concern about drugs is that it is the severity of drug *abuse* rather than the rate of drug *use* that is important. If this is correct, the number of drug-related emergency room visits (arguably the best indicator of the incidence of drug abuse) should correspond to levels of public concern about drugs. In fact, DAWN (Drug Abuse Warning Network) data do indicate that the number of cocaine-related emergency room visits increased between 1986 and 1989, as did public concern (although the increase in public concern was much more uneven). After a brief drop in 1990, however, the number of cocaine- and heroin-related emergency room visits continued to increase. By 1992, the number of all drug emergency room visits—including those involving cocaine and heroin—had reached record levels.[19] By contrast, the percentage of poll respondents identifying drugs as the nation's most important problem in 1992 had dropped from 64% to less than 10%. Neither the reported incidence of drug use nor this measure of drug abuse is consistently related to levels of public concern about drugs.

In sum, from 1964 to 1974, levels of political initiative on and media coverage of crime were significantly associated with subsequent levels of public concern, but the reported incidence of crime was not. From 1985 to 1992, political initiative on the drug issue—but not the reported incidence of drug use or abuse—was strongly associated with subsequent public concern about drugs. These results indicate that the extent to which political elites highlight the crime and drug problems is closely linked to subsequent levels of public concern about them and thus suggest that political initiative played a crucial role in generating public concern about crime and drugs. While it is possible that the measure of public concern used in this analysis fails to capture the extent to which crime and drugs remain of concern even when not seen as the nation's most important problems, this analysis clearly shows that such latent concern is likely to be mobilized and given expression in response to political initiative. In the case of the war on crime, independent media stories also appear to have had an important influence on public perceptions of the crime problem.

The lack of an association between the reported rate of crime and drug use and public concern around those issues is not unique to the two time periods analyzed here. During the middle and late 1970s, for example, reported rates of both crime and drug use increased dramatically: official statistics indicate that the incidence of crime peaked in 1981, while general drug use reached its zenith in 1979 and declined consistently thereafter.[20] Despite this, the percentage of poll respondents identifying crime or drugs as the nation's most important problem remained quite low throughout this period.

The Importance of Political Initiative

The results of the regression analysis presented here cast doubt on the democracy-at-work thesis and document instead a close connection between political initiative and subsequent levels of public concern about crime and drugs. However, there is reason to believe that public concern and political initiative move in similar directions and are mutually reinforcing.[21] Indeed, it is unlikely that political elites—particularly those seeking reelection—would persist in their efforts to mobilize concern about crime and drug use if the public did not appear to be receptive to them.

Public receptivity, however, is not the same as public initiative, and it may be possible to determine whether shifts in levels of political activity precede or follow corresponding shifts in levels of public concern. In order to determine the relationship between these variables over time, those instances in which public opinion shifted most dramatically are presented in diagram form in table 2.3.[22] In each case, the percentage of poll respondents identifying crime (including delinquency and lawlessness/unrest) or drugs as the nation's most important problem appears beneath the poll date; the average number of political initiatives per day in the period between polls appears on the line above those dates.

In each of these cases, public concern and political initiative move largely in parallel directions. In each one, however, a drop in the level of political initiative that is not preceded by a corresponding drop in public concern occurs toward the end of the cycle. For example, in Case 1, public concern about crime, delinquency, and unrest reached its zenith (15%) in October 1968, near the end of an election campaign in which street crime was a central issue. Political initiative was at an all-time high of 1.03 initiatives per day in the period preceding this poll. Nonetheless, the postelection period saw declining levels of political initiative on the crime issue, which were in turn followed by drops in public concern.

Similarly, in Case 3, the percentage of poll respondents reporting that drugs were the nation's most important problem reached its peak at the end of a period of unprecedented political antidrug activity. In

TABLE 2.3. Political Initiative and Public Concern About Crime and Drugs

	Political initiative (above date line) and public concern (below date line)								
Case 1,		.25		.52		1.03		.31	
Crime (January 1968–	1/68	---------->	4/68	---------->	7/68	---------->	10/68	---------->	1/69
January 1969)	8%		10%		13%		15%		12%
Case 2,		.37		.50		.77		.50	
Crime (May 1969–	5/69	---------->	1/70	---------->	5/70	---------->	10/70	---------->	2/71
January 1971)	8%		12%		12%		22%		9%
Case 3,		.38		.53		1.4		.83	
Drugs (September 1988–	9/88	---------->	1/89	---------->	5/89	---------->	9/89	---------->	1/90
December 1989)	15%		11%		27%		64%		33%
Case 4,		.24		.42		1.01		.19	
Drugs (January 1986–	1/86	---------->	4/86	---------->	7/86	---------->	10/86	---------->	1/87
January 1987)	1%		3%		8%		11%		5%

late August and early September of his first year in office, President George Bush made several speeches on "the drug crisis" and called a great deal of attention to his program for fighting drugs. The average number of political initiatives increased from .53 to 1.4 during this period; a public opinion poll administered in late September indicated that 64% of the American public—the highest percentage ever recorded—thought that drugs were the most important problem facing the nation. As in the previous case, subsequent drops in the level of political initiative were followed by declining levels of public concern about drugs. The same pattern is also evident in the other two cases: sudden drops in political attention to the crime and drug issues are not explicable in terms of prior shifts in levels of public concern but are followed by declining levels of public concern.

Similarly, there is no evidence that political elites' initial involvement in the wars on crime and drugs was a response to popular sentiments. Public concern about crime was quite low when candidate Barry Goldwater decided to run on a law and order platform in the 1964 presidential election.[23] Similarly, when President Ronald Reagan first declared a "national war on drugs" in 1982 and when he called for a renewal of this campaign in 1986, fewer than 2% of those polled identified drugs as the nation's most important problem. Nor is the most recent reincarnation of the crime issue a response to popular concern, although the public's attention has certainly shifted in that direction. Only 7% of those polled identified crime as the nation's most important problem in June 1993, just before the legislative debate over anticrime legislation began. Six months later, in response to the high levels of publicity these legislative activities received, that percentage had increased to 30%.[24] By August 1994, a record high of 52% of those polled were most concerned about crime. Gallup Poll analysts concluded that this result was "no doubt a reflection of the emphasis given to that issue by President Clinton since he announced his crime bill in last January's State-of-the-Union Address, and of the extensive media coverage now that the crime bill is being considered by Congress."[25] Ironically, both the UCRs and victimization surveys indicate that the prevalence of most types of crime decreased during this period.

The public's propensity to identify crime and drugs as the nation's most important problems, then, is not primarily shaped by the reported incidence of those phenomena but does appear to be consistently related to prior political initiative on them. But even if the reported incidence of these problems and levels of concern about them were correlated, there is no reason to believe that this anxiety about crime would necessarily lead Americans to identify enhanced punishment as the best response to this problem.

Risk, Concern, Fear, and Support for Punitive Policies

The democracy-at-work thesis rests on the assumption that the risk of criminal victimization increases, anxiety about crime and support for punitive policies will also grow. A wide body of survey research, however, suggests that this set of assumptions is problematic. First, it is not at all clear that one's risk of criminal victimization is consistently related to support for punitive anticrime measures. Despite the fact that rates of criminal victimization are much higher among blacks, for example, it is whites who have historically been more supportive of punitive anticrime measures.[26] Furthermore, while there is some evidence that blacks' experience and fear of victimization in recent years are associated with increasing levels of support for punitive policies, whites' risk and experience of victimization remain unrelated to support for such policies.[27] Individual level data confirm that whites who are at greater risk of victimization are not necessarily more punitive; some studies even report that victims of crime are less punitive than those who have not been victimized.[28] In sum, neither the risk nor actual experience of criminal victimization is consistently correlated with support for punitive policies. White punitiveness in particular seems to be largely inexplicable in terms of one's "risk profile."

Concern about crime and fear of criminal victimization (independent of one's actual risk) also appear to be unrelated to support for tough anticrime measures.[29] It is true that recent increases in concern about crime do correspond to increased punitiveness, but this has not always been the case. During the 1950s, for example, the percentage of people reporting high levels of concern about crime was small but support for punitive anticrime measures was high. Similarly, fear of crime is low but support for tough policies strong among rural and southern white men.[30] Conversely, those most fearful of criminal victimization—blacks and women in particular—are less rather than more supportive of punitive policies. Thus, while those who are at greater risk of victimization (blacks) or are more vulnerable (women and the elderly) do tend to be more anxious about the prospect of being victimized, those who are more fearful are not necessarily more punitive.[31] It is clear that one's risk of or anxiety about criminal victimization cannot explain support for tough anticrime policies.

Crime, Drugs, and the Politics of Representation

To argue that anxiety about crime is not primarily determined by its reported incidence does not imply that the nature and incidence of crime-related problems are entirely irrelevant to public perceptions of them. Indeed, anticrime and drug crusades often rest on a kernel of truth that helps to explain their perpetuation.[32] For example, while the incidence of drug use generally declined in the 1980s, heavy use of cocaine and its derivative, crack, did increase in the mid and late 1980s.[33] The spread of crack—combined with its association with young, nonwhite males, violent crime, and urban blight—undoubtedly facilitated the construction of drug use as the nation's most pressing problem. As noted earlier, however, the continued increase in drug-related emergency room visits after 1990 did not generate high levels of public concern about drugs. Similarly, the politicization of crime in the 1960s was clearly fueled by fears of urban riots and reported increases in the crime rate. But the most recent anticrime campaign (1993–1994) occurred as the reported crime rate plummeted and in the absence of widespread unrest.

Although the relationship between the incidence of crime-related problems and the sociopolitical response to them is complicated, it is clear that popular attitudes about crime and drugs have been shaped to an important extent by the definitional activities of political elites. These actors have drawn attention to crime and drug use and framed them as the consequence of insufficient punishment and control. It is to this sociocultural and quite political process that we may now turn our attention.

3

Creating the Crime Issue

While it is clear that attitudes about crime and punishment are linked to public discourse on those topics, the question remains: how and why were crime-related issues constructed as problems of insufficient punishment and control? Drawing on an analysis of political rhetoric on crime,[1] this chapter traces the emergence and application of this ideological framework and suggests that the discourse of law and order was initially mobilized by southern officials in their effort to discredit the civil rights movement. As the decade progressed, opponents of the welfare state also used this rhetoric to attack President Lyndon Johnson's Great Society programs and the structural explanations of poverty with which they were associated. Discussions of crime were a particularly effective vehicle for promoting the view that poverty and crime are freely chosen by dangerous and undeserving individuals "looking for the easy way out." Somewhat contradictorily, conservatives also identified the "culture of welfare" as an important cause of "social pathologies"—especially crime, delinquency, and drug addiction. Despite their differences, these neoclassical and cultural theories similarly identify "permissiveness" as the cause of crime-related problems and imply the need to adopt policies that would enhance social control rather than social welfare. In short, the creation and construction of the crime issue in the 1950s and 1960s reflect its political utility to conservative opponents of social and racial reform. The following discussion of the allocation of crime control responsibilities in the United States brings into sharp relief the political nature of this appropriation of the crime issue.

Crime Control in American History

The U.S. Constitution allocates most crime control duties to local and state law enforcement. After the Revolutionary War, federal responsibilities were limited to acts that injured or interfered with the federal government. As a result of the growth of interstate commerce and transportation, federal criminal jurisdiction expanded somewhat in the nineteenth century. In the 1920s, bureaucratic efforts to augment the authority of the Federal Bureau of Investigation (FBI) and the prohibition of alcohol further increased the federal government's crime control responsibilities.[2] Despite these modifications, the control of crime remained primarily a state and local responsibility.[3]

The FBI campaign and Prohibition were components of the nation's first war on crime which took place in the context of a crackdown on immigration and political dissent. While there may or may not have been an actual increase in crime during this period,[4] the crime issue became a favorite among politicians. In 1925, President Calvin Coolidge announced the appointment of the first National Crime Commission. This commission accomplished little, but did symbolize the federal government's increased involvement in anticrime efforts. The politicization of crime also had more concrete consequences: between 1917 and 1927 judges delivered significantly longer prison sentences and used the death penalty more frequently.[5] In 1928, Herbert Hoover successfully campaigned on a law and order platform; later that year, the majority of those polled in a national survey felt that crime and disrespect for the law were the nation's most important problems.[6]

In addition to its obvious bureaucratic origins, historians have suggested that this anticrime effort was part of a larger effort to strengthen the position of middle and upper class Americans vis-à-vis the growing immigrant population.[7] Progressive reformers sought to professionalize law enforcement, minimize the power of ethnic ward bosses, alter the ethnic composition of the urban police force, and increase the role of the federal government in anticrime efforts. In addition, immigration and heredity were identified as chief causes of crime during this period; these theories were an important means by which policies limiting immigration to the United States were justified.

Shortly after this controversial appearance on the political scene, the crime issue largely disappeared from national politics. Its disappearance was not complete: in the 1950s, the Kefauver Commission called attention to the dangers of organized crime, Congress passed legislation calling for the use of mandatory sentences for drug offenders, and public concern about juvenile delinquency—prodded by President Harry Truman's attorney general and FBI Director J. Edgar Hoover—increased. But these anticrime initiatives do not compare with the publicity or intensity of the first war on crime. It was not until the 1960s that crime would reemerge as a major issue in national politics. The origins of this development lie in the South and, in particular, in southern officials' attempts to define civil rights protest activities as criminal rather than political in nature.

The Politics of Protest

In the years following the Supreme Court's 1954 *Brown v. Board of Education* decision, civil rights activists across the South used "direct action" tactics in an attempt to force reluctant southern states to desegregate public facilities. Initially, the civil rights movement enjoyed a relatively high degree of public support outside the South. By contrast, southern governors and law enforcement officials characterized its tactics as criminal and suggested that the rise of the civil rights movement was indicative of the breakdown of law and order.[8] Crime rhetoric thus reemerged in political discourse as southern officials called for a crackdown on the "hoodlums," "agitators," "street mobs," and "lawbreakers" who challenged segregation and black disenfranchisement.

As civil rights became a national issue, characterizations of civil rights protests as criminal also became common in national political discourse. For example, after a hesitant President John F. Kennedy finally expressed his willingness to press for the passage of civil rights legislation in 1963, Republicans and southern Democrats criticized Kennedy for "rewarding lawbreakers."[9] Later, a retired Supreme Court justice made the link between crime and protest more explicit when he attributed the spread of lawlessness and violence to

> [t]he fact that some self-appointed Negro leaders who, while professing a philosophy of nonviolence, actually tell large groups of poor and uneducated Negroes . . . whom they have harangued, aroused and inflamed to a high pitch of tensions, that they should go forth and force the whites to grant them their rights.[10]

Justice Charles Whittaker further argued that the current rash of lawlessness and crime was

> fostered and inflamed by the preachments of self-appointed leaders of minority groups . . . [who told their followers] . . . to obey the good laws but to violate the bad ones. . . . This simply advocates the violation of the laws they do not like . . . and the taking of the law into their own hands.[11]

Former Vice President Richard Nixon concurred with this analysis, arguing that "the deterioration [of respect for the rule of law] can be traced directly to the spread of the corrosive doctrine that every citizen possesses an inherent right to decide for himself which laws to obey and when to disobey them."[12]

The Crime Issue in National Politics

Rhetoric regarding the breakdown of law and order appeared more prominently on the national political scene in 1964 when Republican candidate Barry Goldwater announced that "[t]he abuse of law and order in this country is going to be an issue [in this election]—at least I'm going to make it one because I think the responsibility has to start some place."[13] Despite the fact that crime did not even appear on the list of issues considered to be the nation's most important, Goldwater campaigned largely on a law and order platform:

> Tonight there is violence in our streets, corruption in our highest offices, aimlessness among our youth, anxiety among our elderly. . . . Security from domestic violence, no less than from foreign aggression, is the most elementary form and fundamental purpose of any government, and a government that cannot fulfill this purpose is one that cannot command the loyalty of its citizens. History shows us that nothing prepares the way for tyranny more than the failure of public officials to keep the streets safe from bullies and marauders. We Republicans seek a government that attends to its fiscal climate, encouraging a free and a competitive economy and enforcing law and order.[14]

Goldwater promised that, unlike Johnson, he "would not support or invite any American to seek redress . . . through lawlessness, violence, and hurt of his fellow man or damage of his property."[15] Goldwater was not alone in linking opposition to civil rights legislation to calls for law and order: indeed, the most ardent opponents of civil rights and desegregation were also most active on the emerging crime issue. George Wallace, for example, argued that "the same Supreme Court that ordered integration and encouraged civil rights legislation" was now "bending over backwards to help criminals."[16] Three other well-known southern segregationists—Senators James McClellan, Sam Erwin, and Strom Thurmond—led the legislative battle to curb the Supreme Court's efforts to protect the rights of criminal defendants.

Initially, the Republican call for federal leadership in the effort to control crime was controversial among both liberals and conservatives. The proposed federal anticrime effort would not only compete with the Great Society programs for funds, but was inconsistent with the federalist allocation of crime-fighting responsibilities discussed earlier. The idea that the federal government should increase its involvement in the fight against crime therefore contradicted the conservative emphasis on "states' rights" and local responsibility for law enforcement.[17] Because Goldwater and his ilk focused primarily on street crime, this concern was especially relevant.

At the same time that civil rights activists were being identified as enemies of law and order, the FBI was reporting steady increases in the crime rate. Despite significant controversy over their accuracy[18] these reports received a great deal of publicity and were represented as further evidence of the breakdown of morality and lawfulness. The fact that the reported rate of white victimization remained constant during this period was not well publicized, and concern about crime and support for punitive policies increased dramatically among whites.

In sum, the introduction and construction of the crime issue in national political discourse in the 1960s was shaped by the definitional activities of southern officials, presidential candidate Goldwater, and the other conservative politicians who followed his cue. Categories such as street crime and law and order conflated conventional crime and political dissent and were used in an attempt to heighten opposition to the civil rights movement. Conservatives also identified the civil rights movement—and in particular, the philosophy of civil disobedience—as a leading cause of crime. These forms of protest were depicted as criminal rather than political in nature, and the excessive "lenience" of the courts was also identified as a main cause of crime. Countering the trend toward lawlessness, they argued, would require holding criminals—including protesters—accountable for their actions through swift, certain, and severe punishment.

As we will see in chapter 6, the racial subtext of these arguments was not lost on the public: those most opposed to social and racial reform were also most receptive to calls for law and order. Ironically, it was the success of the civil rights movement in discrediting more explicit expressions of racist sentiment that led politicians to attempt to appeal to the public with such "subliminally" racist messages.[19] In subsequent years, conservative politicians also found the crime issue—with its racial subtext now in place—useful in their attempt to discredit welfare programs and their recipients.

Crime, Poverty, and Welfare

Intellectuals and politicians "discovered" poverty in the early 1960s. This "discovery" began with President Kennedy's tour of rural Appalachia in 1961 and the publication of Michael Harrington's best-selling book *The Other America* in 1962. In order to account for the existence of 40 to 50 million poor people, many intellectuals and politicians drew upon Oscar Lewis's formulation of the "culture of poverty," which conceived of poverty as "a way of life . . . passed down from generation to generation along family lines." Insofar as this explanation emphasized the behaviors and values of individuals as a cause of poverty, this theory was "easily appropriated by conservatives in search of a modern academic label for the undeserving poor."[20]

Those who attributed poverty to the lifestyle choices and behaviors of the impoverished often used crime and delinquency to illustrate their argument. For example, Moynihan's now infamous report on the black family attributed black poverty to the "subculture . . . of the American Negro" and the "tangle of pathology" that characterized it. At the heart of this "pathological tangle" was the black family:

> a community that allows large numbers of young men to grow up in broken families, dominated by women, never acquiring any stable relationships to male authority, never acquiring any set of rational expectations about the future—that community asks for and gets chaos. Crime, violence, unrest, disorder, are not only to be expected, but they are very near to inevitable. And they are richly deserved.[21]

The existence of female-headed households, Moynihan argued, resulted in "welfare dependence" and the "failure of youth"—as evidenced by high rates of delinquency, crime, and drug addiction. Although Moynihan (sometimes) identified unemployment as the cause of family "disorganization," subsequent newspaper accounts and conservative reinterpretations of the report did not. Both conservative and liberal culture of poverty formulations thus attributed poverty at least in part to the characteristics and lifestyle choices of the poor. Crime, delinquency, drug abuse, and violence served as highly charged signifiers of this dysfunctionality.

These discussions of the behavioral characteristics of the impoverished were consistent with American officials' long-standing preoccupation with distinguishing the worthy from the unworthy poor and were particularly useful to the conservative effort to emphasize and enlarge the latter category.[22] In the conservative discourse on poverty, the (alleged) misbehaviors of the poor were transformed from adaptations to poverty that had the unfortunate effect of reproducing it into character failings that accounted for their poverty in the first place. The effort to promote cultural explanations of poverty thus led conservatives to highlight the behavioral pathologies and especially the criminality of the poor. Crime-related behaviors—with all their racial connotations and emotional properties—were particularly effective signifiers of the alleged immorality of the poor. The ascendance of this set of images had significant consequences for poverty policy; as Michael Katz suggests, "when the poor seemed menacing they became the underclass."[23]

Those who argued that poverty is a product of personal immorality also insisted that crime and unrest originate in individual choices (shaped by "excessive lenience") rather than social conditions. "How long are we going to abdicate law and order—the backbone of any civilization—in favor of a soft social theory that the man who heaves a brick through your window is simply the misunderstood and underprivileged product of a broken home?" demanded House Leader Gerald Ford.[24] Later, presidential candidate George Wallace also ridiculed "soft social theories" that stress the social causes of crime:

> If a criminal knocks you over the head on your way home from work, he will be out of jail before you're out of the hospital and the policeman who arrested him will be on trial. But some psychologist will say, well, he's not to blame, society is to blame. His father didn't take him to see the Pittsburgh Pirates when he was a little boy.[25]

Rhetoric about crime-related problems was thus used to illustrate the dysfunctional nature of the poor and to promote individualistic explanations of a variety of social problems. With the mobilization of the welfare rights movement, conservatives found crime, delinquency, and drug addiction useful once again—this time as symbols of the moral and familial disintegration caused by the expansion of the welfare state.

Welfare as a Cause of Crime

After the passage of the Civil Rights Act in 1964, civil rights activists and others turned their attention to economic issues and argued that socioeconomic inequality and racism were the main causes of poverty and related social problems. The emergence of the welfare rights movement also led many to agitate for the expansion and transformation of Great Society programs. As a result, the "welfare rolls" grew dramatically: while in 1960 fewer than 600,000 families applied for Aid to Families with Dependent Children (AFDC) benefits, more than three million Americans received such benefits by 1972.[26] Continued migration from southern and rural areas meant that increasing numbers of those who received AFDC were African-Americans.

Theories that attributed poverty to the rejection of the work ethic were easily adapted to this situation, and conservatives soon began to argue that programs such as AFDC encouraged nonwork-oriented lifestyles, thereby reproducing poverty. According to this interpretation, human nature is such that people will avoid work when possible; welfare programs reward this tendency. Public assistance programs also "breed" dependence and stifle initiative in the children of "welfare families": those who are raised by "welfare mothers" fail to learn the values and skills that make productive work likely. The conservative discourse on poverty was (and is) thus gendered in important ways: while the disreputability of welfare mothers centers on charges of sexual promiscuity, male members of the underclass were seen as "lazy or unable to learn the cultural requirements of work and its requirements."[27]

Conservatives began to use the crime issue in their critique of the welfare state as early as 1964. For example, Goldwater argued in the 1964 election campaign that welfare programs are an important cause of increased lawlessness and crime:

> If it is entirely proper for the government to take away from some to give to others, then won't some be led to believe that they can rightfully take from anyone who has more than they? No wonder law and order has broken down, mob violence has engulfed great American cities, and our wives feel unsafe in the streets.[28]

In this twist on the culture of poverty thesis, conservatives argued that the "culture of welfare" undermined self-discipline and promoted "parasitism"—legal (welfare dependency) and illegal (crime). "The chain of reasoning was that crime, civil disorder and other social pathologies exhibited by the poor had their roots in worklessness and family instability, which in turn, had their roots in welfare permissiveness."[29] Moynihan frequently gave expression to this view:

> Among a large and growing lower class, self-reliance, self-discipline and industry are waning; . . . families are more and more matrifocal and atomized; crime and disorder are sharply on the rise. . . . Growing parasit-

> ism, both legal and illegal, is the result; so, also, is violence. (It is a stirring, if generally unrecognized, demonstration of the power of the welfare machine).[30]

The "social pathologies" of the poor (including street crime, drug use, and delinquency) were thus redefined as having their cause in overly generous relief arrangements. By the late 1960s and early 1970s, images of (nonwhite) "welfare cheats" and their dangerous offspring were staples of American political discourse. This imagery is a central component of the New Right's political project, for as Hall argues, the popularization of free-market economics depends upon "the image of the welfare scavenger as folk devil."[31]

The Liberal Response

During and after the 1964 presidential campaign, Lyndon Johnson countered the conservative initiative by stressing that crime control is primarily a local responsibility:

> A visitor coming to America for the first time might have been forgiven for assuming that the President of the United States commanded all the city police departments and that control of the courts was his personal responsibility. The first point that must be made again and again . . . is that crime is a local problem. Its control is a local responsibility. . . . [T]he federal government has little or no power to deal with the problem. . . . Nor should it have.[32]

Members of the Johnson administration also attempted to diminish the impact of conservatives' efforts to heighten concern about crime by suggesting that the escalating crime rate was largely a result of the public's increased willingness to report crimes and officials' ability to keep accurate records of these reports.[33]

Finally, Johnson and other liberals argued that antipoverty programs were, in effect, anticrime programs: "There is something mighty wrong when a candidate for the highest office bemoans violence in the streets but votes against the war on poverty, votes against the Civil Rights Act, and votes against major educational bills that come before him as a legislator."[34] Johnson maintained this position upon ascension to office, insisting that social reforms such as the war on poverty and civil rights legislation would get at the "root causes" of criminal behavior. Initially, then, the Johnson administration stressed the social conditions that generate crime and downplayed the significance of the reported increase in the official crime rate. This emphasis on the root causes of crime was a staple of the postwar liberal discourse on crime and was given expression in a variety of liberal media outlets.[35]

By 1965, apparently in response to reports that the conservative approach to crime was more popular among the electorate,[36] liberals began to change course. Only four months after the election, President Johnson announced the creation of the Commission on Law Enforcement and Administration of Justice and declared in an unprecedented "special message" to Congress on crime that "the present wave of violence and the staggering property losses inflicted upon the nation by crime must be arrested. . . . I hope that 1965 will be regarded as the year when this country began in earnest a thorough and effective war against crime."[37] Johnson presented his newly moderated analysis of the crime problem:

> The problem runs deep and will not yield easy and quick answers. We must identify and eliminate the causes of criminal activity whether they lie in the environment around us or in the nature of individual men. . . . [C]rime will not wait until we pull it up by the roots. We must arrest and reverse the trend toward lawlessness.[38]

Toward that end, Johnson initiated the Law Enforcement Assistance Act of 1965 and the "Safe Streets Bill" in 1967. This legislation entailed greater federal support for local law enforcement and represented a shift away from the view that the most important crime-fighting weapons were civil rights legislation, war on poverty programs, and other policies aimed at promoting inclusion and social reform. While Johnson sometimes reiterated his earlier argument that such policies would help reduce crime, administration officials and other liberal politicians now tempered this argument with the claim that these "long-term" solutions must be balanced by the "short-term" need for increased law enforcement efforts.

This shift parallels growing criticism of social scientific explanations of crime and the rehabilitative ideal in liberal and progressive discourse.[39] While Democratic politicians began to emphasize the need for enhanced law enforcement, progressives launched a fairly thorough critique of rehabilitation that centered on the potential it created for the intrusive, discriminatory, and arbitrary exercise of power. As one analyst of liberal thought on crime and punishment put it, "Previously, the rehabilitative ideology had served as the reference point for criticisms of the prison system; now the ideology was itself the subject of criticism."[40] Across the political spectrum, the rehabilitative project and the discourse of root causes was called into question. These developments undoubtedly made it more difficult for liberals to develop a clear alternative to the conservative approach to crime and may therefore have facilitated the Democratic leap upon the law and order bandwagon.

The Politics of "Law and Order" and the Elections of 1968

In the 1968 presidential campaign, Republican candidate Richard Nixon adopted what criminologists would characterize as a neoclassical approach to crime. Insisting that the real cause of crime is not poverty or unemployment but "insufficient curbs on the appetites or implulses that naturally impel individuals towards criminal activities,"[41] Nixon concluded that the "solution to the crime problem is not the quadrupling of funds for any governmental war on poverty but more convictions."[42] The 1968 Republican party platform concurred with Nixon's critique of liberal "permissiveness": "We must re-establish the principle that men are accountable for what they do, that criminals are responsible for their crime."[43] Independent candidate George Wallace similarly campaigned on his law and order credentials. Only presidential hopeful Hubert Humphrey (quietly) offered a more moderate position on the issue: "We must commit ourselves to make life worth living for every American. Equal protection for all against crime must be a policy not of repression but of liberation; a policy not in reaction to fear but in affirmation of hope."[44]

As a result of its prominence in the election campaign, the crime issue received an unprecedented level of political and media attention in 1968.[45] And the conservative initiative bore fruit: by 1969, 81% of those polled believed that law and order had broken down, and the majority blamed "Negroes who start riots" and "communists" for this state of affairs.[46] Most of the federal anticrime dollars allocated in the late 1960s and early 1970s were spent on police hardware and training programs aimed at containing riots and urban protests.[47] Crime, political dissent, and race were thus merged in both the rhetoric and practice of law and order.

The Federalist Dilemma

Upon ascension to office, the Nixon administration was forced to contend with the fact that the federal government has little authority to deal directly with street crime outside of Washington, D.C. A dismayed Attorney General John Mitchell pointed out that "even if the federal government found an indirect way of intervening in the problem, the local government would get the credit for diminishing those classes of crime."[48] White House aide Egil Krogh agreed:

> The President had campaigned on his desire to reduce crime, to reduce crime nationwide. Crime had to be stopped. I don't think as a matter of intelligent politics he could have been in office for one year and then said, "I've discovered that the federal government has little jurisdiction over street crime . . . and therefore it is a matter for the states to handle. Good luck!"[49]

Insiders concluded that "the only thing we could do was to exercise vigorous symbolic leadership" and therefore waged war on crime by adopting "tough sounding rhetoric" and pressing for largely ineffectual but highly symbolic legislation.[50] As one White House aide admitted, "[W]hile these bills would suggest a tough law and order demeanor by the Administration, the legislation itself did not provide an enhanced ability to the police departments or to the courts to reduce crime."[51] Journalists began to report that despite Nixon's tough talk, the crime rate was still rising.

Administration officials attempted to resolve this dilemma in several ways. First, as we will see in chapter 7, federal aid to local and state law enforcement increased dramatically during this period. The administration's High Impact Anti-Crime Program, for example, targeted mid-sized cities with Law Enforcement Assistance and Administration (LEAA) discretionary funds. In addition, new statistical artifacts were created in the hope that these would permit a more flattering assessment of Nixon's capacities as a crime fighter. One of the more notorious of these was created to show that the *rate of increase* in the crime index was decreasing.[52] Most important, however, was the administration's identification of narcotics control—for which the federal government has significant responsibility[53]—as a crucial anticrime weapon.[54] The resulting war against drug abuse thus emerged as a last-ditch attempt to reduce the crime rates to which the administration itself had drawn so much attention.[55]

In order to explain and legitimate this new strategy, administration officials argued that drug addicts commit the majority of street crimes in order to pay for their drugs. The adoption of this line of reasoning helps to account for the Nixon administration's somewhat incongruous support for methadone maintenance programs, designed to reduce the likelihood that addicts would steal to finance their habit. The Nixon administration continued its wars on crime and drugs until the outbreak of the Watergate scandal in 1974, an event that diverted the nation's attention from "crime in the streets" to "crime in the suites."

The Discourse of Law and Order in Historical Context

While the conservative discourse on crime created some dilemmas for the national politicians who promoted it, the issue was, for the most part, politically and ideologically useful. Crime-related problems served as an especially effective vehicle for reconstructing popular conceptions of the poor. As we will see in chapter 6, receptivity to this discourse depended to a significant extent on its racial connotations. A more complete understanding of the larger political and electoral context helps to explain how racialized imagery and language came to serve as such an important resource for national politicians.

Partisan Dealignment and the Southern Strategy

The New Deal coalition—an alliance of urban ethnic groups and the white South—dominated electoral politics from 1932 to the early 1960s. As a result of black migration to the North, this alliance included more and more blacks—a trend that marked a dramatic break from the post–Civil War partisan configuration and created quite a dilemma for those interested in maintaining white southern allegiance to the Democratic party. In 1948, President Harry Truman responded to the increasing number of black voters by pressing for a relatively strong civil rights platform and the first serious signs of strain in the Democratic partnership appeared. White southerners organized a "states' rights" party, and in the subsequent election, four deep-South states (Louisiana, South Carolina, Alabama, and Mississippi) delivered their electoral votes to this insurgent political force. In the 1952 and 1956 elections, Democrats attempted to placate the "Dixiecrat" delegates and pull in disaffected southerners. But the appeasement of southern racism was not without political costs, as many northern blacks defected from the Democratic party and the Republican share of the black vote increased from 21% in 1952 to 39% in 1956.[56] The continuing migration of many southern blacks to northern cities meant that blacks constituted an increasing portion of the total vote.

In 1957 and 1960, partisan competition for the black vote led the Democratic Congress to pass the first civil rights measures of the twentieth century. Convinced he could not resurrect southern loyalty to the Democratic party, Kennedy campaigned on a civil rights platform in 1960. Once in office, however, he sought to minimize southern resistance within the Democratic coalition; this ambivalence about the loss of the white South appears to account for his weak and delayed support for civil rights legislation. Indeed, it was only under the extreme pressure generated by civil rights activists that Kennedy declared his allegiance to the civil rights cause.

It was thus the civil rights movement that finally cut the South from the Democrats and enabled the GOP to make a bid for that region. As conservative political analyst Kevin Phillips wrote in 1969, "The Negro problem, having become a national rather than a local one, is the principal cause of the break-up of the New Deal Coalition."[57] By drawing significant public attention to the plight of blacks in the South, civil rights activists forced the national Democratic party to choose between its southern white and northern black constituencies. The high degree of support among nonsouthern whites for the civil rights cause prior to 1965 and the increasing numbers of northern black voters eventually led the Democratic party to cast its lot with blacks and their sympathizers.

This decision, however, alienated many of those traditionally loyal to the Democratic party, particularly southerners. "Millions of voters, pried loose from their habitual loyalty to the Democratic party, were now a volatile force, surging through the electoral system without the channeling restraints of party attachment."[58] These voters were "available for courting," and courted they were. As early as 1961, Goldwater argued that "we [the Republicans] are not going to get the Negroes as a block in the '64 or '68 elections, so we might as well go hunting where the ducks are."[59] Initially, the GOP targeted white southerners—voters who had formerly made up the "solid South." This strategy was quite successful: analyses of the 1964 elections indicate that the socioeconomic class structure of the New Deal alliance could be fractured by the issue of race. In the poorest white neighborhoods of Birmingham, for example, the Republican vote increased from 49% to 76%, and a similar trend could be discerned in other southern cities.[60] This approach became known as the "southern strategy" and was quite successful in attracting white southerners to the ranks of the GOP.

Republican analysts suggested that they might also find a responsive audience among white suburbanites, ethnic Catholics in the Northeast and mid-West, blue-collar workers, and union members. Patrick Buchanan declared that "a New Majority," including the traditional Republican political base, the solid South, the farm vote, and half the Catholic, blue-collar vote of the big cities, could dominate electoral politics.[61] Some conservative political strategists frankly admitted that appealing to racial fears and antagonisms was central to this strategy. For example, Phillips argued that a Republican victory and long-term realignment was possible primarily on the basis of racial issues and therefore suggested the use of coded antiblack campaign rhetoric (e.g., law and order rhetoric).[62] Similarly, John Ehrlichmann, special counsel to the president, described the Nixon administration's campaign strategy of 1968: "We'll go after the racists. That subliminal appeal to the anti-black voter was always present in Nixon's statements and speeches. . . ."[63] As the traditional working-class coalition that buttressed the Democratic party was ruptured along racial lines, race eclipsed class as the organizing principle of American politics. By 1972, attitudes on racial issues rather than socioeconomic status were the primary determinant of voters' political self-identification.[64]

The New Right

The movements for racial and social reform not only precipitated this dealignment and "racialization" of the American party system, but also engendered a powerful backlash on the right. The most successful current within this backlash is known as the New Right, so named in order to be distinguished from the traditional, East Coast leadership of the Republican party. The New Right differentiated itself from the Old Right in two important ways. First, the New Right adopted a more populist stance and "rejected the view that unrestrained expressions of popular will militate against the orderly processes of government on which stable societies depend."[65] Second, the New Right wed traditional conservative economic policies and anticommunism to a conservative stance on contemporary "social issues," especially those with racial implications. New sets of constituencies were mobilized through the use of racially charged "code words"—phrases and symbols that "refer indirectly to racial themes but do not directly challenge popular democratic or egalitarian ideals."[66] The law and order discourse is an excellent example of such coded language, and allowed for the indirect expression of racially charged fears and antagonisms.

The emphasis on social issues such as crime were thus part of the New Right's attempt to secure consensus around a conservative set of political interpretations and policies. To a certain extent this effort has been guided by electoral considerations: the New Right, based primarily in the Republican party, has rearticulated racial meanings in such a way as to encourage defections from the Democratic party. This strategy enabled the Republican party to replace the New Deal cleavage between the "haves" and the "have-nots" with a new division between some (mostly white) working and middle class voters and the traditional Republican elite, on the one hand, and "liberal elites" and the poor on the other.[67] But the New Right's "authoritarian-populist"[68] project is aimed, more broadly, at discrediting state policies and programs aimed at minimizing racial, class, and gender inequality and strengthening those that promise to enhance the states' control of the troublesome. Law and order rhetoric has been a particularly important means by which conservative elites attempted to justify this reconstruction of the state's role and responsibilities, and the racialization of American politics created fertile soil for the creation and mobilization of the crime issue.[69] The following chapter documents the continuation of this ideological campaign and liberals' increasing acceptance of its main assumptions.

4

From Crime to Drugs—and Back Again

The salience of the crime and drug issues declined dramatically following President Richard Nixon's departure from office. Neither President Gerald Ford nor President Jimmy Carter mentioned crime-related issues in their State of the Union addresses or took much legislative action on those issues. During and after the 1980 election campaign, however, the crime issue once again assumed a central place on the national political agenda. Like conservatives before him, candidate and President Ronald Reagan paid particular attention to the problem of street crime and promised to enhance the federal government's role in combating it. Once in office, however, the institutional difficulties associated with this project led the Reagan administration to shift its attention from street crime to street drugs. Political and public concern about the drug problem increased throughout the 1980s; by August 1989 President George Bush characterized drug use as "the most pressing problem facing the nation." Shortly thereafter, a New York Times/CBS News Poll reported that 64% of those polled—the highest percentage ever recorded—thought that drugs were the most significant problem in the United States.[1]

Like crime, drug use was defined in political discourse as a social control rather than a public health or socioeconomic problem. And as the decade progressed, the public became more likely to support enhanced law enforcement efforts, harsher sentences, and the contraction of civil rights as appropriate solutions to the problem of drugs.[2] This shift was part of a more general trend toward toughness that began in the 1960s (although, as we will see in subsequent chapters, there is reason to suspect that this shift is more superficial than is commonly supposed). This time, however, not only conservatives played a leading role in the campaign to get tough: many Democratic policymakers attempted to wrest control of the crime and drug issues from the Republicans by advocating stricter anticrime and antidrug laws.

Interest in the drug issue faded following the outbreak of the Persian Gulf War in 1991. In 1993, however, the crime issue was resuscitated yet again by both conservative and ostensibly liberal policymakers. The result has been an unprecedented bipartisan consensus regarding the need to expand the size, scope, and resources of the crime control apparatus and—not coincidentally—to "end welfare as we know it."

This chapter analyzes the development of the war on drugs in the 1980s and the reappearance of the crime issue in the early 1990s, and shows that national political discourse on these issues continues to be profoundly shaped by the original conservative framework described in the previous chapter. Although the framing of the crime and drug issues has not changed, the extent of the involvement—even initiative—of leading Democratic officials in the campaign to get tough is new. In exercising this initiative, liberals and conservatives alike draw on a rich cultural legacy in which discussions of crime and drugs often serve as vehicles for the construction of the poor as an undeserving and "dangerous class."

Creating the Dangerous Classes

The notion that crime and poverty have their roots in the lifestyles and preferences of the poor has a long history in American political culture; racial and ethnic stereotypes have often informed this conception of poverty-related problems. Popular discourse on drug use has also developed in such a way as to reinforce the image of the poor as morally depraved. During what Reinarman and Levine call "drug scares," moral entrepreneurs blame a variety of social problems on chemical substances and those who imbibe them.[3] Temperance advocates, for example, attributed many of the social ills associated with modern industrial society—crime, vice, poverty, disease, and the breakdown of the family—to the consumption of alcohol. In this and other antidrug crusades, racist imagery and the association of drug use with crime generated widespread fear, even panic, about chemical substances.[4]

The first law restricting the use of opium, for example, was adopted in California in the 1870s in the context of an economic depression for which Chinese immigrants were blamed. As the economic situation worsened, the practice of smoking opium (but not taking it in pill form, as most Anglo-Americans did) was prohibited. Similarly, the first cocaine scare occurred in the post-Reconstruction South, where the image of a coke-crazed black man led some police departments to switch from 32- to 38-caliber guns (despite the absence of evidence indicating that the consumption of cocaine was at all common among southern blacks). The first major campaign against marijuana occurred in the Southwest during the Depression, as unemployment rates skyrocketed and Mexicans—associated with the use of marijuana—were blamed for increasing unemployment rates and declining standards of living. In this context, marijuana was described as the "killer weed," likely to make people—and particularly Mexicans—more violent. The association between drugs, crime, and racialized images of dangerous classes, then, has characterized antidrug crusades throughout American history.[5]

The relationship between drug use and crime has been the subject of a tremendous amount of social scientific research, the results of which are somewhat ambiguous. Drug abuse and criminality coexist in some social groups but not in others. Among those populations where such a relationship does exist, criminality tends to precede drug use rather than vice versa.[6] More recently, researchers have found that much of the association between violence and drugs is a product of the illegal nature of the drug market and the socioeconomic context in which battles over market share are fought.[7] Despite this empirical complexity, the political and ideological connection between drugs, crime, and dangerous classes has remained intact. The historical development of the Reagan/Bush war on drugs was informed by and reinforced this connection.

Poverty, Welfare, and the Revival of the Crime Issue

As governor of California and presidential candidate, Ronald Reagan consistently emphasized the problem of street crime and the need for a more punitive approach to it. Shortly after his election, Reagan announced his proposed anticrime package and reasserted the fundamentals of the conservative position on crime:

> We can begin by acknowledging some absolute truths. . . . Two of those truths are: men are basically good but prone to evil; some men are very prone to evil—and society has the right to be protected from them. . . . [T]he war on crime will only be won when an attitude of mind and a change of heart takes place in America—when certain truths take hold again . . . truths like: right and wrong matters; individuals are responsible for their actions; retribution should be swift and sure for those who prey on the innocent.[8]

Reagan's portrait of the criminal ("a stark, staring face—a face that belongs to a frightening reality of our time: the face of the human predator . . . . Nothing in nature is more cruel or more dangerous. . . .")[9] was certainly intended to bolster support for such retributive endeavors.

Once Reagan was in office, his attorney general, William French Smith, appointed the Attorney General's Task Force on Violent Crime to recommend "ways in which the federal government can do more to combat violent crime."[10] Because state and local governments are largely responsible for identifying and prosecuting conventional street crime, however, the administration's desire to involve the federal government in combating violent crime was problematic. The Reagan administration nonetheless began to pressure federal law enforcement agencies to set aside their focus on white-collar offenses and shift their attention to street crime instead. As Evelle Younger, chairman of Reagan's Advisory Group on the Administration of Justice, said, "most of us want to focus on violent crime, crime in the streets. . . ."[11] Similarly, Donald Santorelli, advisor to the Reagan transition team, criticized the Carter administration's "preoccupation with white-collar crime." By October 1981, the Justice Department released a report announcing its intention to cut in half the number of specialists assigned to identify and prosecute white-collar criminals. But not just white-collar crime was excluded from the Reagan administration's crackdown on crime: according to David Davis, former staff member of the Attorney General's Task Force on Violent Crime, many subjects that could have been considered components of the "violent crime" problem—including corporate violence—were never mentioned in the task force's discussions of that issue. Similarly, domestic violence was explicitly rejected as "not the kind of street violence about which the Task Force was organized."[12]

As discussed in the previous chapter, this focus on street crime—with all its racial connotations—is partially explicable in terms of the Republican party's electoral strategy. Richard Wirthlin, director of the Reagan/Bush planning committee, described the Reagan campaign strategy this way: "Right from the beginning, it was recognized that we had to solidify the Republican base and broaden it. As our key swing targets we selected ethnic Catholics, labor, blue-collar workers, and we felt that we could make a major run at the Carter coalition in the South."[13] The conventional wisdom among the Republicans was that the way to attract working-class men and their families was on the basis of what Reagan called the social issues—especially law and order. Consistent with this analysis, the Republican party platform of 1980 advocated "firm and speedy application of criminal penalties," increased use of the death penalty, and the "firm punishment of drug pushers and drug smugglers with mandatory sentences."[14]

The Reagan and Bush administrations' emphasis on street crime was also part of the ongoing effort to generate support for conservative economic and social policies. The get-tough discourse continued to provide an ideal opportunity to espouse the view that human vice and greed were the cause of social problems, to criticize the "liberals" who wrongly blamed "society" for them, and, somewhat contradictorily, to attribute crime, delinquency, and drug abuse to the welfare programs and "lenient" crime policies of these well-intentioned but misguided liberals. Each of these applications of crime discourse is described in the following sections.

Crime and Human Nature

Like conservatives before them, the Reagan and Bush administrations went to great lengths to reject the notion that street crime and other social problems have socioeconomic causes. Reagan's first major address on crime, for example, consisted of a sweeping philosophical attack on "the social thinkers of the fifties and sixties who discussed crime only in the context of disadvantaged childhoods and poverty-stricken neighborhoods."[15] This theme appeared again and again in Reagan's speeches on crime:

> Here in the richest nation in the world, where more crime is committed than in any other nation, we are told that the answer to this problem is to reduce our poverty. This isn't the answer. . . . *Government's function is to protect society from the criminal, not the other way around*[16] (my emphasis).

As both vice president and president, George Bush also criticized the notion that crime is related to its social context:

> We must raise our voices to correct an insidious tendency—the tendency to blame crime on society rather than the criminal. . . . I, like most Americans, believe that we can start building a safer society by first agreeing that society itself doesn't cause the crime—criminals cause the crime.[17]

Conservatives also argued that the erroneous belief that crime is the product of social arrangements simply allows criminals to avoid responsibility for their actions:

> . . . [I]t is abundantly clear that much of our crime problem was provoked by a social philosophy that saw man as primarily a creature of his material environment. The same liberal philosophy that saw an era of prosperity and virtue ushered in by changing man's environment through massive Federal spending programs also viewed criminals as the unfortunate products of poor socio-economic conditions or an underprivileged upbringing. Society, not the individual, they said, was at fault for criminal wrongdoing. We were to blame. Well, today, a new political consensus utterly rejects this point of view. . . .[18]

According to Reagan, then, "the American people have lost patience with liberal leniency and pseudointellectual apologies for crime."[19] This new "political consensus" emphasized choice:

> Choosing a career in crime is not the result of poverty or of an unhappy childhood or of a misunderstood adolescence; it is the result of a conscious, willful choice made by some who consider themselves above the law, who seek to exploit the hard work and, sometimes, the very lives of their fellow citizens.[20]

Furthermore, the reality of human nature is such that only the threat of punishment will deter criminal behavior.

> The crime epidemic threat has spread throughout our country, and it's no uncontrollable disease, much less an irreversible tide. Nor is it some inevitable sociological phenomenon. . . . It is, instead, and in large measure, a cumulative result of too much emphasis on the protection of the rights of the accused and too little concern for our government's responsibility to protect the lives, homes, and rights of our law-abiding citizens. . . . [T]he criminal element now calculates that crime really does pay.[21]

In accordance with this neoclassical philosophy, Reagan eliminated TASC (Treatment Alternatives to Street Crime) and stressed instead the need for enhanced law enforcement and punishment, aimed at raising the costs of "choosing" evil.

Crime and the Welfare State

According to the interpretation outlined above, the naive view that social inequality is criminogenic led liberals to believe that the war on poverty would help reduce crime. Conservatives complained that these "social thinkers, with their Utopian presumptions about human nature, [have] hindered the swift administration of justice [and] also helped fuel the expansion of government."[22] In criticizing these "Utopian ideas," these critics depicted programs that provide public assistance to the poor as examples of "excessive lenience." By contrast, conservatives suggested that

> Americans object to government intrusion into areas where government is neither competent nor needed, but . . . they [are also] critical of government's failure to perform its legitimate and constitutional duties like providing for the common defense and preserving domestic tranquility.[23]

Lest they be perceived as mean-spirited, these critics of the welfare state depicted public assistance programs as detrimental to the poor themselves and conservatives as the true allies of the impoverished:

> By nearly every measure, the position of poor Americans worsened under the leadership of our opponents. Teenage drug use, out-of-wedlock births, and crime increased dramatically. Urban neighborhoods and schools deteriorated. Those whom the government intended to help discovered a cycle of dependency that could not be broken. Government became a drug, providing temporary relief, but addiction as well.[24]

Crime is one of the main consequences of the assistance programs foisted upon the poor by liberals:

> In the welfare culture, the breakdown of the family, the most basic support system, has reached crisis proportions—in female and child poverty, child abandonment, horrible crimes. . . .[25]

Conservatives thus argued that welfare programs such as AFDC not only "keep the poor poor," but also accounted, along with lenient crime policies, for the rising crime rate. In this discourse, "generous welfare provisions and soft criminal justice policies are entwined in their detrimental effect upon morality and responsibility for the increasing crime problem."[26] This argument was an attempt to legitimate reductions in welfare spending as well as the implementation of increasingly punitive crime and drug policies:

> Our current welfare program, originally designed to raise people out of poverty, has become a crippling poverty trap, destroying families and condemning generations to a dependency. . . . Of course, one of the best things we can do for families is obliterate drug use in America. . . . [We must therefore make] society intolerant to drug use with stiff penalties and sure and swift punishment for offenders.[27]

In sum, the conservatives suggested that government's functions had been distorted: the state would be on more legitimate constitutional grounds and would more effectively "help the poor" by scaling back public assistance programs and expanding the criminal justice system and law enforcement:

> [T]his is precisely what we're trying to do to the bloated Federal Government today: remove it from interfering in areas where it doesn't belong, but at the same time strengthen its ability to perform its constitutional and legitimate functions. . . . In the area of public order and law enforcement, for example, we're reversing a dangerous trend of the last decade. While crime was steadily increasing, the Federal commitment in terms of personnel was steadily shrinking. . . .[28]

Reagan thus articulated the central premise of the conservative project of state reconstruction: public assistance is an "illegitimate" state function, whereas policing and social control constitute its real "constitutional" obligation. The conservative mobilization of crime-related issues was thus a component of the effort to reconstruct popular images of the poor and thereby legitimate the contraction of public assistance programs and the expansion of the social control apparatus.

Because motives are notoriously difficult to ascertain, it is not possible to determine whether the use of crime-related issues toward this end was consciously strategic. Some statements by Republican party strategists, however, do suggest such intent. Lee Atwater, for example, explained the GOP strategy in the following manner:

> There are always newspaper stories about some millionaire that has five Cadillacs and hasn't paid taxes since 1974. . . . And then they'll have another set of stories about some guy sitting around in a big den saying so-and-so uses food stamps to fill his den with booze and drugs. So it's which one of these that the public sees as the bad guy that determines who wins. . . .[29]

While some of the more astute political players may have consciously wielded the crime-drug-welfare issue in an effort to shape perceptions of "the bad guy," it is also quite likely that this set of interpretations is experienced as "truth" by many of its proponents. Given the experiences, vantage point, and goals of its advocates, the crime discourse just described undoubtedly appears natural or common-sensical.[30]

Political rhetoric notwithstanding, the view that crime had its origins in humankind's propensity to evil or in welfare "dependence" was not supported by a new political consensus. Throughout the late 1970s and early 1980s, most Americans continued to attribute crime to socioeconomic conditions. In 1981, for example, a national poll found that unemployment was most likely to be identified as the main cause of crime. Similarly, an ABC News Poll taken in 1982 found that 58% of those polled named unemployment and poverty as the most important causes of crime; only 12% identified "lenient courts" as the main source of this problem.[31] Over time, however, the public did become more likely to reject these structural explanations. By 1989, 60% would report that cutting the drug supply was the most effective anticrime measure, while only 10% would identify reducing unemployment as the most important means of fighting crime.

From Crime to Drugs

Realizing the commitment to reducing street crime through a tough law enforcement approach was complicated by the fact that fighting street crime is primarily the responsibility of local law enforcement. As a result, FBI Director William Webster initially resisted suggestions that his agency shift its resources to fighting this type of crime, arguing that fighting street crime "is not our role, it's not our responsibility."[32] To bolster his case, Webster cited a study which found that "[P]eople consider bank embezzlements more serious than many thefts and burglaries, a bribe of $10,000 to a legislator more serious than a $100,000 bank burglary, and a retail price-fixing scheme more serious than a robbery where an armed subject intimidated a victim and took $1,000."[33]

One month later, however, Webster announced that "the drug problem has become so widespread that the FBI must assume a larger role in attacking the problem." In explaining this shift, Webster argued that "when we attack the drug problem head on, it seems to me that we are going to make a major dent in attacking violent street crime. . . ."[34] It seems that Webster had accepted the proposal that the FBI shift its focus to street crime; narcotics control—which falls partly under the jurisdiction of federal law enforcement—provided the mechanism by which federal law enforcement could become more involved in the war on street crime.

As a result—and in stark contrast to agencies with drug education, prevention, and treatment responsibilities—federal law enforcement was able to stave off the General Accounting Office's (GAO) proposed "across-the-board" budget cuts. For example, FBI antidrug monies increased from $8 million in 1980 to $95 million, and the budget of the Drug Enforcement Agency (DEA) increased from $215 to $321 million between 1980 and 1984. Antidrug funds allocated to the Department of Defense more than doubled (from $33 to $79 million) during this period, while the Customs Department's allocation grew from $81 to $278 million.[35] The budgets of these law enforcement agencies increased at even more rapid rates in the years ahead: total federal expenditures for law enforcement activities increased from $2.2 billion in 1980 to $5.6 billion in 1991.[36] In contrast, funding for agencies with responsibility for drug treatment, prevention, and education was sharply curtailed. The budget of the National Institute on Drug Abuse, for example, was reduced from $274 million to $57 million between 1981 and 1984, and antidrug funds allocated to the Department of Education were cut from $14 million to $3 million. By 1985, 78% of the funds allocated to the drug problem went to law enforcement, while only 22% went to drug treatment and prevention.[37]

It is clear, then, that the Reagan administration's early emphasis on street crime and the need for a punitive approach to it gave a distinct advantage to law enforcement agencies in the bureaucratic scramble for antidrug funds. White House Counselor Edwin Meese, for example, played an important role in "encouraging" Webster to shift the focus of the FBI to street crime and drug trafficking. Similarly, over half of the recommendations of the Attorney General's Task Force on Violent Crime (appointed by the Reagan administration) pertained to the control of narcotics. Thus, while federal bureaucracies attempted to enhance their resources by emphasizing their antidrug capacities, federal law enforcement agencies were encouraged to do so by administration officials themselves.

Not all members of the Reagan administration were equally enthusiastic about increasing the responsibility (and budgets) of federal law enforcement, however. The Office of Management and Budget (OMB), for example, had a very different agenda. When OMB Director David Stockman advocated cutting the federal law enforcement budget in March 1981, he was met with fierce opposition from Attorney General Smith. As Stockman laments:

> Attorney General William French Smith did not think his department was a place to start economizing. "The Justice Department is not a domestic agency" he said. "It is the internal arm of the nation's defense. . . ." . . . If anything, he said, the Reagan administration would have to spend more on law enforcement, rather than less. . . . [O]nce the Attorney General had christened his agency an "Internal Defense Department" we would have lots of law enforcement at the federal level, even if we couldn't afford it. Justice's budget would grow and grow as the Attorney General came up with more and more schemes to show that the administration was "committed" to aggressive "internal defense."[38]

As Stockman indicates, Reagan sided with law enforcement in this dispute, reiterating the notion that social control (as opposed to social welfare) is a true governmental responsibility: "Bill is right. Law enforcement is something we have always said was a legitimate function of government."[39] As a result, most federal law enforcement agencies were excepted from the Reagan administration's proposed cuts for federal agencies.

In September 1982, Attorney General Smith, seeking to further expand the budget of the Justice Department, began to argue that it was imperative that an official war on drugs be declared. The OMB argued that this new assault should be delayed for budgetary reasons. President Reagan once again sided with the Justice Department, and in October 1982 President Reagan officially announced his administration's "war on drugs." In pressing for greater resources, heads of law enforcement agencies appealed to Congress, claiming that their efforts to join the battle against crime and drugs were being thwarted by the administration (and the OMB in particular). Legislators concerned about being perceived as "soft on crime" and members of subcommittees whose authority is linked to the fate of particular federal law enforcement agencies were especially anxious that the Reagan administration fulfill its commitment to the battle against crime and drugs.

For example, in 1981, Customs Commissioner William von Raab began to lobby for technologically advanced equipment to increase Customs' ability to halt drug smuggling. When White House budget cutters refused, von Raab appealed to Congress. By 1988, Congress had allocated an additional $700 million dollars for this purpose, and Customs' arsenal had increased from two intercept aircraft to 88 planes and helicopters (despite the fact that the agency itself estimated that only 20% of the cocaine entering the country enters on airplanes). Representative Glenn English of Oklahoma, chair of a key subcommittee, was crucial in rallying support for von Raab's campaign, and Customs soon began construction of a new center for the coordination of its air efforts in land-locked Oklahoma City, English's home district.[40] The Democratic tendency to argue that the war on drugs was underfunded (rather than ill-conceived) thus emerged early in the 1980s.

In sum, the administration's emphasis on the need for a tough approach to crime facilitated the emergence of the war on drugs and shaped the nature of that campaign.[41] While the Reagan administration always placed great emphasis on the importance of law enforcement and punishing drug offenders, most of this rhetoric was aimed at the "drug pushers" and "narco-traffickers" who "preyed on our young people" prior to 1986. After this time, however, the antidrug campaign was enlarged to also include casual drug users: "A new understanding is evident: Drug abuse is not a private matter. Using illegal drugs is unacceptable behavior. And the costs are paid by all of society."[42] Furthermore,

> If this problem is to be solved, drug users can no longer excuse themselves by blaming society. As individuals, they're responsible. The rest of us must be clear that . . . we will no longer tolerate the illegal use of drugs by anyone.[43]

This banner was valiantly carried forward under President George Bush:

> What's the difference, then, between the wonderful young kids behind me, this great-looking group back there, and the kids who huddle a few blocks from where we stand, using and dealing drugs? Same schools. Same Houston—but a different choice.[44]

Public Opinion and the War on Drugs

While the impetus for the war on drugs in the 1980s came from within the federal government, public opinion has not been irrelevant to the development of federal drug policy. In fact, public support for the war on drugs has played an important role in legitimating the expansion and intensification of the antidrug campaign. But the argument that the Reagan administration "harnessed a preexisting momentum for a crackdown on drugs"[45] is not supported by the available evidence: public opinion polls indicate that public concern about drugs did not increase prior to the Reagan administration's declaration of war in 1982. For example, as of 1981, only 3% of the American public believed that cutting the drug supply was the most important thing that could be done to reduce crime, while 22% felt that reducing unemployment would be more effective. Furthermore, the percentage of poll respondents identifying drug abuse as the nation's most important problem had dropped from 20% in 1973 to 2% in 1974 and hovered between 0% and 2% until 1982. In sum, there is no evidence of an upsurge in concern about drugs prior to Reagan's declaration of war.[46] The erroneous identification of public opinion as the primary impetus for the government's campaigns against crime and drugs obscures the political nature of those efforts.

The Escalation of the War on Drugs

Political and media concern about the drug issue intensified in the summer of 1986. While much of this publicity centered on the cocaine-related deaths of athletes Len Bias and Don Rogers, a variety of other factors also contributed to the escalation of antidrug rhetoric and efforts.[47]

In October 1985, the DEA sent Robert Stutman to serve as director of its New York City office. Stutman made a concerted effort to improve relations with the news media and sought to draw journalists' attention to the spread of crack cocaine. "The agents would hear me give hundreds of presentations to the media as I attempted to call attention to the drug scourge . . . ," Stutman wrote later. "I wasted no time in pointing out its [the DEA's] new accomplishments against the drug traffickers and using those cases to illustrate the full scope of the drug abuse problem."[48] Stutman explains his strategy as follows:

> In order to convince Washington, I needed to make it [drugs] a national issue and quickly. I began a lobbying effort and I used the media. The media were only too willing to cooperate, because as far as the New York media was concerned, crack was the hottest combat reporting story to come along since the end of the Vietnam war.[49]

This media campaign appears to have been quite effective: the number of drug-related stories appearing in the *New York Times* increased from 43 in the latter half of 1985 to 92 and 220 in the first and second halves of 1986, and the number of drug-related stories published in the *Times* in these years was far greater than the number published in other newspapers.[50] The success of Stutman's campaign is not surprising: as Fishman and others have pointed out, news workers' primary source of information about crime is law enforcement.[51]

The administration's previous antidrug efforts also contributed to the media's identification of drug use as a major news story. While waging its war against drugs, the administration and other government agencies had disseminated a great deal of antidrug "information." These claimsmaking activities facilitated the identification of drug (and especially crack) use as a news "theme." The availability of such themes has important implications for media coverage: "crime incidents are rarely reported unless news workers see them as related to past or emerging trends in criminality or law enforcement."[52]

Increased coverage in the *New York Times* had quite significant consequences as other media outlets soon followed suit. In June 1986, for example, *Newsweek* declared crack to be the biggest story since Vietnam/Watergate, and in August of that year *Time Magazine* termed crack "the issue of the year." The number of television network news stories focusing on drugs increased from 73 in the second half of 1985 to 103 and 283 in the first and second halves of 1986. This evidence of "intermedia influence" is consistent with a wide body of research that demonstrates the tendency of media to look to other media outlets—especially the *New York Times*—for confirmation of their news judgment.[53] Much of the drug-related news coverage during this period emphasized the spread of crack-related violence to white communities, the threat of random (drug-induced) violence to which this "epidemic" gave rise, and the need for enhanced surveillance and policing in order to establish control over the burgeoning crack trade and the violence it spawned.[54] Because these stories highlighted the threat of random violence, they appear to have contributed to growing support for a quick and dramatic response to the drug problem.[55]

In an attempt to ensure that their party received credit for taking action on the emerging drug issue, Democrats in the House of Representatives began putting together legislation calling for increased anti-drug spending. This activity on Capitol Hill triggered an even higher degree of media interest in the drug issue: while less than 1% of all news coverage focused on drugs in the early 1980s, that percentage increased to 3.2% in July 1986 and to 6% in the two-week period ending August 10.[56] During August and September, the television networks continued to allocate a tremendous amount of news time to the drug issue and offered specials such as "48 Hours on Crack Street." It was in this context that President Reagan made several nationally televised speeches on the drug problem. The first of these speeches, made on August 4, 1986, was typical in tone and content: Reagan emphasized that drug users must be held accountable for their use of drugs: "drug users can no longer excuse themselves by blaming society. As individuals, they are responsible. The rest of us must be clear that we will no longer tolerate drug use by anyone."[57]

Many politicians claimed that their initiative on the drug issue was a response to growing public concern about drugs. President Reagan, for example, asserted that "the polls show that this [drugs] is, in most people's minds, the number one problem in the country."[58] A reporter attempted to identify the source of this claim at a White House Press briefing with President Reagan's spokesman, Larry Speakes:

> QUESTION: The President recently cited a poll in which he said that 71%, I believe, of the American public cited drugs as the number one issue. Do you know what poll that was . . . ?
>
> SPEAKES: I don't know—sure don't. Bill?
>
> QUESTION: Larry, if I could continue—you said that there has been a tremendous outpouring of public feeling since the Len Bias death. Do you have any research or evidence of what kind of public feeling there is on this issue?

> SPEAKES: No. I just think it's an obvious feeling about the amount of publicity that was given to the most recent sports drug deaths that have really peaked [sic] public interest. . . .[59]

Media interest in the drug issue (which was itself related to official antidrug activity) was thus interpreted as a sign of growing public concern about drugs. Indeed, congressional Republicans warned Reagan that unless he came up with more specific antidrug proposals quickly they would be compelled to endorse the $2 to $3 billion "alternative" promoted by the Democratic leadership. And so they were: on September 12, the House passed legislation that allocated $2 billion to the antidrug crusade for 1987, required the participation of the military in narcotics control efforts, allowed the death penalty for some drug-related crimes, and allowed for the admission of some illegally obtained evidence in drug trials. Later that month the Senate proposed even tougher antidrug legislation, and on October 28, the president signed the Anti-Drug Abuse Act of 1986 into law.

In fact, national polls administered prior to this spate of legislative activity do not indicate that public concern about drug use had grown significantly. According to a *New York Times*/CBS News Poll taken in April 1986, only 3% of those polled were most concerned about drugs. By late August, however, after the publicity surrounding Reagan's antidrug speeches and legislative activity on the issue, a *New York Times*/CBS poll reported that the percentage of Americans who felt that drugs were the most important national problem had increased to 13%.[60] Thus, by the time legislation was created, debated, and signed into law, the polls did show that more Americans were more concerned about drugs (although concern about economic issues was far greater).

In the period from 1986 to 1990, drug use was frequently one of the nation's most publicized issues. Public concern about drugs reached its zenith immediately following President Bush's national address in 1989 in which he focused exclusively on the drug crisis. Under Bush, federal funds allocated to the battle against drugs were greater than under all presidents since Richard Nixon combined, and a record 3.5 million drug arrests were made during this period. The crime issue also enjoyed a high profile in the late 1980s, as exemplified by Bush's successful manipulation of what came to be known as the "Willie Horton incident" in the 1988 elections[61] and conservatives' continued condemnation of those who have "become lost in the thickets of liberal sociology."[62] The neoclassical and cultural theories discussed earlier characterized discussions of the crime and drug issues throughout this period. In fact, it was after 1985 that the alleged behavioral attributes (including crime, drug use, and violence) of the "underclass" received the most publicity.[63]

Crime and drugs retained their prominence on the national political agenda through 1990, but the outbreak of the Persian Gulf War precipitated their decline and President Bush largely ignored these topics during the campaign season of 1992. This shift probably reflects the failure of the war on drugs (as indicated by the continued increase in drug-related emergency room visits, only marginal declines in overall drug use, and an increasing supply of cocaine and heroin within the United States), as well as candidate Bill Clinton's relative invulnerability on these issues. Like many "new" Democrats, Governor Clinton was quite determined not to suffer the fate of Democratic presidential candidate Michael Dukakis, who was portrayed by the Bush administration as hopelessly "soft on crime." As governor and presidential candidate, Clinton expressed strong support for expanded police efforts, more aggressive border interdiction programs, and tougher penalities for drug offenders. As a result, "[T]here was little about Clinton's crime control record in Arkansas that Bush could taunt him about the way he mocked Dukakis as a patsy for every dark-skinned murderer in Massachusetts."[64] The 1992 Democratic platform also embraced the idea that levels of crime and drug use are a direct function of crime control efforts: "The simplest and most direct way to restore order in our cities is to put more police on the streets."[65] It was in the context of such bipartisan consensus that the crime issue made its most recent comeback.

Return of the Crime Issue

Despite his record as governor and his relatively tough talk during the election campaign, there was some speculation that the election of Bill Clinton would faciliate a diminution of the get-tough approach to the crime and drug problems. His record and campaign rhetoric were somewhat ambiguous in this regard: on the one hand, Clinton emphasized the need for greater law enforcement efforts as well as boot camps for juvenile offenders and touted his record on capital punishment. On the other hand, both before and after the election, Clinton occasionally evinced hints of an alternative analysis of the crime problem. For example, in a speech to the Democratic Leadership Council shortly after the Los Angeles uprising, Clinton characterized looters as people whose "lives and bond to the larger community had been shredded by the hard knife of experience." Clinton also criticized the Reagan-Bush administrations for blaming crime problems on "them"—poor, nonwhite Americans. And Clinton had spoken eloquently of the need to reverse the trend toward racial isolation and the government's responsibility to redress existing inequalities. A year after the election, Clinton still, at least occasionally, gave expression to these views:

> We have to rebuild families and communities in this country. We've got to take more responsibility for these little kids before they grow up and start shooting each other. I know the budget is tight, but I'm telling you, we have to deal with family, community and education, and find jobs for members of society's underclass to bring structure to their lives.[66]

In short, Clinton sometimes espoused the notion that crime is related to social conditions and some hoped that his election might therefore create a space for developing and implementing an alternative approach to crime.

This potential was not realized. In August 1993, Republicans announced an anticrime legislative package calling for more police, enhanced federal support for prison construction, and limits on habeas corpus appeals. One week later, Clinton and several key congressional Democrats proposed their own anticrime legislation calling for more police, enhanced federal support for prison construction, and limits on habeas corpus appeals. The only significant difference between the two parties' approaches to crime control was the issue of gun control,[67] and this was the subject of much debate throughout 1993. This pattern was reinforced in November when the results of several key elections were interpreted as expressions of the public's desire to get tough with criminals.[68] In December, the Senate passed legislation authorizing federal funds to help state governments hire more police officers and build prisons and expanded the death penalty for dozens of new federal crimes. The publicity associated with this legislation appears to have had an impact on public opinion: the percentage of those polled who felt that crime was the nation's most important problem increased from 9% in June 1993 to 22% in October and again to 32% by January 1994.[69]

The publicity surrounding the crime issue intensified when President Clinton used his 1994 State of the Union address to urge further congressional action, including adoption of a federal equivalent of California's "three-strikes" law. Most Democrats—pleased with new poll results indicating that Republicans no longer enjoyed an advantage on the crime issue[70]—continued to support the expansion of law enforcement and the criminal justice system while offering only tepid criticism of some mandatory sentencing provisions and mild support for some preventive measures.[71] The $30 billion crime bill sent to President Clinton in August 1994 was hailed as a victory for the Democrats who "were able to wrest the crime issue from the Republicans and make it their own."[72] The final legislation authorized $8.8 billion for hiring more police and $7.9 billion in state prison grants, created dozens of new federal capital crimes, and mandated life sentences for some three-time offenders.

Less than one-fourth of the funds appropriated by this legislation were earmarked for preventive measures. Even so, Republicans in Congress attacked the inclusion of this "pork spending" and opposed the Democratic crime legislation on this basis. Some time later, these and other newly elected lawmakers pledged in their Contract With America to replace the rather meager funding for social programs with more flexible block grants to state and local officials. This and other goals—especially the hope of easing the restrictions on gun ownership that had been adopted—were embodied in a series of bills passed by the House in February 1995. This legislation also increased federal grants for state prison construction by another $2.6 billion, restricted the scope of court-ordered settlements in prison conditions lawsuits, limited inmates' rights to sue over these issues, expanded federal prosecutors' capacity to use evidence obtained illegally, further restricted prisoners' ability to file habeas corpus petitions, and strengthened measures to provide for the swift deportation of illegal immigrants.[73] Asked to explain President Clinton's failure to provide any real alternative to these proposals, one administration official said, "[Y]ou can't appear soft on crime when crime hysteria is sweeping the country. Maybe the national temper will change, and maybe, if it does, we'll do it right later."[74]

What this official failed to grasp was the role that Clinton and other political elites played in the generation of the public concern to which they were subsequently compelled to respond. Throughout the 1980s, Democratic party officials had increasingly made the conservative rhetoric on crime and drugs their own. As one analyst concluded, "[T]his "purging of 'root causes' and rehabilitation from the crime debate can be traced to the 'law and order' policies of the late 1960's. . . . [I]t was ultimately the Reagan-Bush 'war on drugs' and crack cocaine in the 1980's that kept crime and violence a major public issue, with the crimes and drugs of minorities and the economic underclass at the center of the crime agenda."[75] The liberal about-face on crime-related problems reflected, among other things, conservatives' ability to disseminate law and order rhetoric through the mass media, as well as its apparent resonance with electorally important segments of the American public. These, then, are the subjects of the following chapters.

5

Crime and Drugs in the News

For people living in modern society, the media are a crucial source of information about the social and political world. This is particularly true with respect to crime: over 90% of those polled report that the media is their most important source of information about this social problem.[1] Research investigating the effects of this reliance indicates that the media play an important "agenda-setting" function: "the press may not be successful in telling people what to think, but they are stunningly successful in telling people what to think about."[2] Specifically, these studies report that those social issues and problems that receive a high degree of attention in the news are more likely to be identified as the nation's most important by the viewing public.[3]

More recent research indicates that the content of media products also has an important impact on the formation of political opinions. For example, Iyengar found that stories characterized by "episodic" news frames (in which public issues are discussed in terms of specific instances) are much more likely to elicit individualistic attributions of responsibility than are "thematic" stories that place public issues in a more general and historical context.[4] Surveys show that heavy consumers of violent television crime shows are more likely to see the world as a violent and frightening place and to adopt a "retributive justice perspective."[5] Similarly, experimental studies indicate that those who are exposed to media discussions of serious (i.e., violent) crimes are more likely to subsequently perceive other crimes as more serious and to support punitive anticrime measures.[6]

It appears, then, that both the quantity and nature of media imagery influence members of the public as they make assessments of social issues and their solutions. Because officials are privileged in the battle for media exposure, these actors also play an important role in the formation of political opinion and in the creation and regulation of culture.

The State, The Media, and the Social Construction of Reality

The idea that the state plays an influential role in the cultural and social-psychological processes described above runs counter to popular conceptions of the democratic process. According to this perspective, the institutionalization of democratic citizenship in the United States entailed the development of formal mechanisms (such as electoral institutions and legislative bodies) for the consultation and expression of public opinion. Indeed, the legitimacy of our governing bodies rests on the notion that leaders respond to the "will of the people" as expressed through formal political channels.

While this perspective has some merit, the modern democratic emphasis on public opinion did not lead governments to simply surrender to that opinion as they found it.[7] Instead, the birth of democracy also led to the development of techniques aimed at creating, shaping, channeling, and mobilizing public opinion. Several important historical developments—including the decline of the party as a mass-based organization and the development of broadcast media technologies—have facilitated the government's efforts to shape and harness mass opinion.

The Transformation of Electoral Politics

During the late nineteenth century in the United States the corrupt but assimilative urban machine became an important feature of the political landscape. Party loyalty was quite high during this period and participation of eligible voters (over 80% of all adult males) reached its peak. After the turn of the century, however, Progressive-era reformers saw public opinion as irrational and potentially disruptive and were therefore concerned about the ascendance of the mass party. Their efforts to establish a direct primary system and voting registration requirements, weaken the urban machine, and reduce the frequency of elections were largely successful.[8] The net effect of these reforms was the decline of the party as a mass-based organization, decreasing partisanship, and the disappearance of the partisan press. The public relations profession, similarly born of concern over unruly expressions of mass opinion, also emerged during this period. Together, these developments altered the nature of the electoral campaign: electoral politics increasingly consisted of attempts to appeal to and shape the sentiments of nonpartisan voters through public relations and advertising efforts.[9]

This shift away from mass mobilization and toward advertising in political campaigns coincided with the birth of modern journalism. The emergence of broadcast (wireless) communication technologies in the early twentieth century was facilitated by the joint efforts of the U.S. government and several large corporations. The utility of these efforts to the U.S. government during World War I stimulated continuing investment in broadcast research and development. While it has long been observed that states tell "noble lies" in order to promote their authority, the fact that states could now communicate their claims to the general public marked a dramatic change. As Guglielmo Marconi, the primary architect of the first wireless technologies, put it: "For the first time in the history of the world, man is now able to appeal by means of direct speech to millions of his fellows, and there is nothing to prevent an appeal being made to fifty millions of men and women at the same time."[10]

The spread of public relations, and especially its use by government officials during and after World War I, had important consequences for news-gathering practices. While journalists in the nineteenth century aspired to report just "the facts" as they saw them, twentieth century reporters increasingly reprinted those "facts" promoted by government officials and others who could afford to buy public relations services. Schudson suggests that the rise of public relations meant that journalists "needed a framework within which they could take their own work seriously. . . . This is what the notion of 'objectivity' . . . tried to provide."[11] The resulting norm of "objectivity" was operationally defined as "the reprinting of consensually validated and authoritative statements."

"Objective" news is thus biased in favor of the definitions of the powerful, and particularly those of state officials: "[T]he slant of [objective] journalism lay not in explicit bias but in the social structure of newsgathering which reinforced official viewpoints of social reality."[12] Given the United States' growing international prominence and the centralization of political functions in Washington, the majority of these "authoritative" statements came from federal officials. The geographic concentration of state sources and their ability to supply frequent and conveniently formatted "news" meant that the use of state sources also satisfied the organizational needs of news workers. As Schudson concludes, "Newspapers that had once fought the interests now depended upon them for handouts."[13]

Officials' public relations efforts are increasingly sophisticated and continue to serve as the basis of many news stories. For example, official news releases are likely to be published with only cursory checks on their veracity.[14] There is also evidence that officials frequently serve as news sources: A recent study found that 72% of all sources for network television news were government officials or leaders of political groups and institutions.[15]As one researcher concluded, "while resistance to direct forms of control has hardened in the press, susceptibility to news management has spread. . . . The routines and conventions of reporters' work incline them to accept the words of the officials without probing beneath them on their own."[16]

It is important to note that the embeddedness of state and media institutions does not mean that officials are universally capable of shaping media representations. Social movement activists, for example, have sometimes been able to use the media to challenge dominant perspectives and disseminate alternative issue frames.[17] Interelite conflict also complicates the ability of one group of state actors to shape media interpretations of social issues. Thus, the struggle to shape media representations is a contest, but one that is played on a quite unequal playing field. The following analysis examines the extent to which official sources were able to frame national crime news between 1964 and 1974 and drug-related news stories between 1985 and 1992.[18]

Framing Crime and Drugs

Social issues may be represented or framed in a number of ways. The discursive elements that make up different issue frames are organized into what Gamson calls "interpretive packages" that make sense of and give meaning to social issues such as crime. At the center of each package is a core frame—a central organizing idea that gives meaning to a series of events or phenomena related to the issue in question. Packages are further characterized by a set of "signature elements" that suggest the core frame and serve as "condensing symbols" for the entire package. It is important to note that these packages are descriptions of pure types; they do not typically appear in media products in their entirety, but are conceptual frames for analyzing the often mixed content and meaning of media products. The "culturally available" crime and drug issue packages are described in the following sections.

Identifying the Crime and Drug Issue Packages

In order to identify all culturally available packages (including those that do not enjoy a great deal of coverage in the mass media), a wide range of publications covering crime-related isues was analyzed. On the basis of this analysis, four crime and drug issue packages were identified. The core frame for each of these packages and the "reasoning devices" that justify it are described here. These devices include a causal analysis of the issue in question, an assessment of the consequence of particular policies, and appeals to principle. The more evocative rhetorical devices used to suggest the different issue frames —exemplars (events that illustrate a particular point), catchphrases (thematic statements or slogans that suggest a particular frame), and depictions (characterizations of principal subjects)—are also described. These elements make up the "signature matrix"[19] of each package and are presented in tables 5.1 and 5.2.

Crime Issue Packages

Respect For Authority Respect for authority has broken down because individuals are not being held responsible for their behavior. The failure to hold people accountable for their actions has its roots in the misguided liberal notion that behavior is "caused": "Our forebears were never concerned about why a person misbehaves. We are straying away from the principle of holding the individual responsible for his actions."[20] The belief that behavior has "causes" has led to the adoption of lenient policies, which have in turn undermined respect for authority and increased crime: "An attitude of permissiveness is becoming more and more evident in our society today, leading to the progressive relaxing and discarding of all forms of restraint and discipline."[21] The approval of "so-called civil disobedience," the increasingly lax judicial system, and permissive welfare arrangements all contribute to the decline in respect for authority. The solution is "to make respect for law and order the first priority in our national life"[22] by enhancing law enforcement and cracking down on criminals.

Stories about people who flagrantly violate the law without fear of punishment and who seek "something for nothing" serve as exemplars in this package. Catchphrases include "mollycoddling," "permissiveness," "accountability," "parasitism," and "law and order." Those who seek to identify and ameliorate the "root causes" of crime are depicted as "bleeding hearts" and "mollycoddlers." This package was sponsored by Republican candidate Barry Goldwater, key congressional Republicans following the 1964 election, and other conservative politicians and jurists.

Balance Needs The central issue in this package is the need to respond to the fear of crime while simultaneously addressing its causes. Fear of crime is a rational response to the increasing crime rate; the public needs to be protected from the threat of criminal victimization. In the short term, therefore, law enforcement efforts must be enhanced and improved so that people can live safe and secure lives. An exclusive focus on the causes of crime would lead us to neglect the importance of maintaining law enforcement. At the same time, long-term solutions, aimed at addressing the deeper causes of crime, are also needed. A balance must be struck: "[T]here is no conflict between this need [for day-to-day law enforcement] and the parallel need to attack the causes of crime. . . . [T]he two needs are complementary. An obsessive emphasis on either . . . can only hamper effective law enforcement."[23]

"Balance" is the main catchphrase in this package; those who focus exclusively on either law enforcement or the causes of crime are depicted as ideological rather than pragmatic. This package also emphasizes the need for enhanced research and technical assistance in order to improve the efficiency of the criminal justice system. *Balance Needs* was the main Democratic alternative to *Respect for Authority* after 1965.

Civil Liberties under Attack The core issue in this package is the need to develop crime policies that are consistent with the principles of democracy and the protection of civil liberties. Law and order policies lead to a disregard for civil rights and due process. To the extent that this is the case, it becomes difficult (if not impossible) to distinguish law breakers from law enforcers. Crime policies that are consistent with the preservation of civil liberties and the due process revolution must be implemented.

Incidents in which the civil rights of the accused have been violated serve as exemplars for this package. Catchphrases include "due process," "civil rights," and "the rights of the accused." This package was sponsored by lawyers and civil rights organizations such as the American Civil Liberties Union (ACLU), some congressional liberals, and other supporters of the Warren Court's due process revolution.

TABLE 5.1. Signature Matrix of Crime Issue Packages

Package	Frame	Position	Exemplars
Respect for authority	The issue is how to instill respect for authority and fear of punishment so that the human propensity to choose evil can be controlled.	Weakening respect for authority and diminishing fear of punishment account for the breakdown of law and order. The leniency of the courts and permissive attitudes and policies have encouraged this trend.	Stories about people flagrantly disregarding or disrespecting authority, obviously guilty persons who are "slapped on the wrist," and people who exploit "lenient" policies to their benefit.
Balance needs	The issue is how to balance our short-term need for law enforcement with the long-term objective of ameliorating the root causes of crime.	Crime policy must seek to identify the root causes of crime and simultaneously satisfy the need for increased law enforcement, particularly among the poor.	Stories that depict the need for policies which address poverty and other social problems as well as people's need for protection.
Civil liberties under attack	The issue is how to address the crime problem in a way that does not undermine civil rights and the due process revolution.	"Tough" anticrime policies often violate our constitutional liberties and rights.	Stories in which the rights of the accused are flagrantly violated.
Poverty causes crime	The issue is the need to ameliorate the "root causes" of crime.	Crime is largely the consequence of social conditions such as poverty, joblessness, and racism.	Stories about people whose life experiences and limited opportunities led them to adopt a life of crime.

Catchphrases	Depictions	Roots	Principles
Respect for authority. Law and order. Handcuffed police. Technicalities. Accountability. Permissiveness. Propensity to evil. Bleeding hearts. Mollycoddling.	Liberals are depicted as mollycoddlers; bleeding hearts; soft-headed.	The root of the problem is the spread of "soft" theories which provide excuses for criminals and other miscreants.	Respect for authority and fear of punishment are the basis of civilized society. Individuals must be held accountable for their actions.
Short- and long-term crime policies must be balanced. Efficiency. Pragmatism.	Those who advocate only increased law enforcement or addressing the root causes of crime are depicted as partisan and ideological rather than pragmatic.	The root of the problem is simplistic and partisan thinking about crime and its solutions.	We must not allow partisan and simplistic thinking to interfere with the creation and implementation of an effective crime policy and an efficient criminal justice system.
Civil rights. Democratic procedure. Due process. Constitutional protections.	Those implementing crime policies are depicted as overzealous and willing to compromise on important civil rights.	The root of the problem is the lack of commitment to libertarian values and individual rights.	Our constitutional guarantees regarding due process and protection of the rights of the accused should be our most important consideration.
Root causes. Structural conditions. Blocked opportunities. The war on poverty is a war on crime.	Conservatives are depicted as more interested in punishing the poor than in reducing crime.	The root of the problem is our unwillingness to take political steps to reduce the conditions that cause crime.	We have a moral obligation to attack poverty and racism rather than punish its victims.

TABLE 5.2. Signature Matrix of Drug Issue Packages

Package	Frame	Position	Exemplars
Get the traffickers	The issue is how to stop the traffickers from preying on young people and teach our kids to "just say no."	Traffickers are an especially vicious group of people and must be stopped at all costs. Drug education is also crucial.	Stories of vicious and heinous acts committed by drug traffickers and pushers; stories about kids who resist peer pressure.
Zero tolerance	The issue is how to reduce the casual drug use that is responsible for the blood lost in the war on drugs.	Casual drug users must be held accountable for their decision to use drugs. We must severely punish all those who violate drug laws.	Stories of people whose lives have been hurt by people's use of drugs; stories of people who use drugs but are not punished.
Need more resources	The issue is how to ensure that politicians live up to their commitment to fighting drugs.	Politicians exploit the drug issue for political gain. They must be pressured to fully fund the war on drugs.	Stories about the administration's unwillingness to commit funds.
War fails	The issue is whether we increase or reduce harm by policies that criminalize and punish drug users.	The harm caused by drugs is increased by making them illegal. A law enforcement approach to the problem fails to address to the "root causes" of drug abuse and actually makes these problems worse.	Stories about individuals serving harsh sentences for minor drug law violations or whose civil rights have been violated in the war on drugs.

Catchphrases	Depictions	Roots	Principles
Narco-traffickers. Drug Kingpins. Merchants of Death. Just say No.	People who sell drugs are depicted as greedy and violence-loving individuals. Kids are depicted as vulnerable and subject to peer pressure.	The root of the problem is the traffickers' use of violence to obtain profits.	The traffickers have no right ro prey on our children. They must be stopped.
Zero tolerance. Casual users must be punished.	Liberals are depicted as criminal accomplices; drug users as criminals.	The root of the problem is the liberal belief that casual drug is okay, and the fact that such users have not been held accountable.	Casual drug users must be held accountable for the damage they cause.
Election year politics. A serious war on drugs will require more funds.	Politicians and especially administration officials who took the initiative on drugs are depicted as insincere and politically motivated.	The root of the problem is the lack of a real commitment to fighting drugs.	Politicians should not use an issue as serious as drug abuse for political reasons. Fighting drugs requires a serious commitment.
Drug warriors. Root causes. Civil rights. Harm reduction.	Those waging the war on drugs are depicted as more interested in punishing people than in reducing the harm associated with drugs.	The root of the problem is our tendency to define problems such as drug use as a criminal problem.	If we are really interested in reducing the harm associated with the use of drugs, we should help rather than punish those for whom it is a problem.

Poverty Causes Crime The core issue is the need to attack the structural causes of crime. "Unemployment, ignorance, disease, filth, poor housing, congestion, discrimination—all of these things contribute to the great crime wave that is sweeping through our nation."[24] People are affected by their environment, and crime is a response to the hopelessness of poverty and racism. If we are serious about solving the crime problem we will need to develop policies that reduce socioeconomic inequality. Policies that do not address these root causes of crime will be ineffective and constitute further punishment for those already most hurt by poverty and racism. The primary catchphrase of this frame is "root cause." This package was sponsored by President Lyndon Johnson and other liberal politicians prior to 1965, a variety of social movement activists, and writers of progressive journals such as *The Nation*.

Drug Issue Packages

Get the Traffickers The core issue in this package is the need to prevent "narco-traffickers" and drug pushers from terrorizing our nation's citizens, especially our children. Law enforcement, border patrol, and the judicial system must be beefed up so that we can more effectively prosecute those traffickers who prey on our kids: "[W]e need new laws to stop drug traffickers from harming our people, especially our young people."[25] Traffickers and pushers are a vicious group of people, and the root of the problem is their greed and violence. These people need to be made aware that we will come after them—that "they can run but they can't hide." Casual users, and particularly children, are their victims and need to be encouraged to resist peer pressure and "just say no." Catchphrases include "merchants of death," "narco-traffickers," "drug kingpins," "pushers," and "just say no." This package was sponsored by the Reagan administration and First Lady Nancy Reagan, some law enforcement programs, programs such as DARE (Drug Abuse Resistance Education), and some parents' organizations.

Zero Tolerance Casual drug users are not victims but criminals: "Drug abuse is not a so-called victimless crime. . . . [T]he victims of this terrible crime . . . are countless. They're the people beaten and robbed by junkies. They're the people who pay higher insurance rates because of such robberies. And they're the people who pay higher prices for goods of all kinds because drugs in the workplace have undermined worker productivity. The victims of drug abuse, in short, are you and me, our friends, our families—all Americans."[26] The central issue in this package is the need to deter drug users and dealers through the threat of increased punishment.

The root of the drug problem is that people have not been held accountable for their use of drugs. Liberal policies that have failed to do so breed addiction and have led to an increase in drug use and crime. Users must pay the price if they break the law. They must be held accountable for the damage they do to society. The criminal nature of the decision to use drugs means that "a massive wave of arrests is a top priority for the war on drugs."[27] Catchphrases include "user accountability," "zero tolerance," and "get tough." This package was sponsored by the Reagan and Bush administrations and law enforcement personnel, particularly after 1986.

Need More Resources The central issue is whether the government will commit sufficient resources to the war on drugs. Because "there's no down side to the drug issue," politicians appear to be focusing their attention on this problem, particularly around election time. But lawmakers must be pressured to allocate resources (especially for treatment and prevention)—not just rhetoric—if the antidrug campaign is going to work. The root of the problem is the insincerity of politicians who are using this issue for political advantage. Without a real commitment to the war on drugs, the problem of drug abuse will worsen. Politicians are depicted as opportunistic and "election-year politics" is the main catchphrase. *Need More Resources* was the main Democratic alternative to the packages sponsored by the Reagan and Bush administrations: some Democratic leaders (and heads of agencies with responsibility for dealing with drug use) attempted to undercut Republican "ownership" of the drug issue by suggesting that the Republican initiative was politically motivated. Medical experts, parents' organizations, and news workers also suggested that interest in the drug issue was politically motivated and highlighted the need for a "real" commitment to the war on drugs.

War Fails The central issue is whether the prohibition of drugs and tough law enforcement lessen or increase the harm caused by drugs. The law enforcement approach threatens important civil rights and actually increases the harm caused by drugs: prohibition creates a black market characterized by high levels of violence and thus exacerbates the plight of poor communities in which this violence is most likely to occur. We need drug policies which recognize that locking people up does not address the fundamental social and economic ("root") causes of drug abuse and distribution. If we are really interested in minimizing the harm caused by drugs, drugs must be treated as a socioeconomic or public health problem.

Catchphrases include "root causes," "harm reduction," and "civil rights," while those who espouse a law enforcement approach to the drug problem are depicted as "drug warriors." The prohibition of alcohol, which also led to increased violence and crime, is an exemplar in this package. This package was sponsored primarily by civil rights organizations such as the ACLU, reform organizations such as the Drug Policy Foundation, and some community activists.

Measuring Package Prominence and Sponsorship

With the signature elements of the crime and drug issue packages identified, the prominence of each package in media discourse may be assessed. By identifying the sponsor of each package display, the extent to which the various issue frames were sponsored by state and nonstate actors was also evaluated. Because of the large number of crime- and drug-related stories appearing during the two periods studied, only those media items that appeared during four sampling periods were analyzed.[28] In the crime case, stories from the *New York Times*, *Washington Post*, and *Los Angeles Times* and indexed under "crime in the United States" were included in the analysis.[29] The sample in the drug case consisted of news stories from the evening broadcast of network television indexed under "drug abuse" or "drug trafficking."[30]

The analysis proceeded as follows. First, displays of any of the signature elements of the various issue packages were identified in each story and then coded according to which package they signified (each media item may display signature elements associated with more than one package). Next, these displays were classified as "state-sponsored" if they were directly attributed to a source currently affiliated with the federal state[31] and as "nonstate-sponsored" if their source was not so affiliated.[32] Only displays that were explicitly attributed to state sources were considered to be state-sponsored; because journalists do not always identify their sources, this undoubtedly underestimates the extent to which state actors influence media coverage.

Crime and Drug Issue Frames in the News

Overall, well over half of the signature element displays were sponsored by state officials. Table 5.3 depicts the number of crime- and drug-

TABLE 5.3. Number and Percentage of Stories and Package Displays in State- and Nonstate-Sponsored Stories

	State	Nonstate	Total
Crime-related newspaper stories	NA	NA	100% (186)
Crime package displays	65% (192)	35% (105)	100% (297)
Drug-related television stories	NA	NA	100% (127)
Drug package displays	76% (201)	24% (64)	100% (265)

related news stories and the number and percentage of state- and nonstate-sponsored package displays they contained.

Given journalists' reliance upon state sources, it is not surprising that media coverage of the crime and drug issues peaked when state activity on these issues was at its highest level. For example, the *New York Times* carried 373 stories on crime in 1968 alone and these stories were particularly concentrated around the elections. Similarly, network television news carried 520 stories about drug use in 1989—the year of President George Bush's most intensive antidrug campaign.

Newspaper Coverage of the Crime Issue

The fact that journalists relied so heavily on official sponsors had significant consquences for the depiction of the crime issue. As Table 5.4 indicates, over three-fourths of the package displays sponsored by state sources depicted *Respect for Authority*, while just 34% of the nonstate-sponsored package displays depicted this frame. The preponderance of state sources meant that elements of *Respect for Authority* were depicted in more than 62% of the total package displays. These stories generally lamented liberal permissiveness and emphasized the need to instill the fear of punishment in order to counter the trend toward lawlessness.

TABLE 5.4. Number and Percentage of State- and Nonstate-Sponsored Crime Package Displays in Newspaper Stories, 1965–1973

Package displays	Respect for authority	Balance needs	Civil liberties under attack	Poverty causes crime	Total displays
State-sponsored	77% (147)	9% (18)	8% (15)	6% (12)	100% (192)
Nonstate-sponsored	34% (36)	10% (11)	15% (16)	40% (42)	100% (105)
Total displays	62% (183)	10% (29)	11% (31)	18% (54)	100% (297)

Elements of *Poverty Causes Crime* were depicted in only 6% of state-sponsored displays but in 40% of those statements offered by nonstate sources. Journalists' reliance upon official sources thus also helps to explain the relative absence of this alternative package in crime news. *Balance Needs* was slightly more popular among state officials while *Civil Liberties under Attack* was more popular among nonstate sources. In sum, the dominance of state sponsors in crime-related news stories (and the ability of conservatives to define themselves as the owners of the crime issue) largely explains the near hegemony of the *Respect for Authority* package in the news.

The Drug Issue in Television News

Television news coverage of drug-related topics in the 1980s was also quite likely to rely on state sources: 76% of all package displays appearing in these stories were attributed to state actors. Because they were especially likely to be sponsored by officials, the two law and order packages—*Get the Traffickers* and *Zero Tolerance*—dominated discussions of the drug issue (see Table 5.5). Many of the stories that depicted these frames focused on "drug busts" and relied on footage provided to them by various law enforcement agencies. Others carried the statements, speeches, and policy proposals of government officials who espoused views associated with these issue packages.

Stories that relied heavily on official sources were also less likely to contain elements associated with *War Fails*: only 4% of the package displays offered by state elites depicted this frame, while nearly a third of the displays sponsored by nonstate actors did so. Overall, the overwhelming presence of state sources in television news coverage of the drug issue meant that this more critical perspective was largely absent from drug-related news stories. *Need More Resources* was more likely to be sponsored by nonstate actors but did not have too significant a presence in drug-related news items. In sum, stories that relied on state sources were more likely to depict get-tough frames and rela-

TABLE 5.5. Number and Percentage of State- and Nonstate-Sponsored Drug Package Displays in Television Stories, 1982–1991

Package displays	Get the traffickers	Zero tolerance	Need more resources	War fails	Total displays
State-sponsored	57% (115)	31% (62)	8% (16)	4% (8)	100% (201)
Nonstate-sponsored	38% (24)	6% (4)	25% (16)	31% (20)	100% (64)
Total displays	52% (139)	25% (66)	12% (32)	11% (28)	100% (265)

tively unlikely to give expression to the ideational elements associated with *War Fails*.[33] The fact that the media relied on the state to the extent that it did therefore had significant consequences for the framing of the drug problem.

Two main conclusions can be drawn from this analysis of crime- and drug-related news coverage. First, crime- and drug-related news stories drew heavily on official sources. Second, officials were able to promote favored issue frames through the mass media and thereby affect the framing of the crime and drug issues in the news. Specifically, stories that relied on official sources were more likely to apprehend the problem of crime in terms associated with the *Respect for Authority* package and to depict the drug issue through the lens of *Get the Traffickers* or *Zero Tolerance*. Journalists' tendency to rely on state sources also accounts for the absence of more critical discussions of the crime and drug issues. The ascendance of the discourse of law and order in the news, then, was largely a consequence of officials' capacity to call attention to and frame discussions of the crime and drug issues. Conservative politicians and law enforcement personnel were particularly successful in defining themselves as the relevant "authorities" on the crime and drug issues.

But news stories are, in the final analysis, written by journalists working for media organizations. Why did news workers fall into lockstep with the conservatives lamenting declining respect for authority and the need to crack down on those involved with drugs? While their reliance on official sources was crucial, journalists' collusion also reflects the fact that covering the crime issue and especially the "drug crisis" from the point of view of law-enforcement agencies (and especially film footage of drug "raids") satisfies the media's interest in dramatic and sensationalistic news. The ride-along footage of drug busts, the touring of enemy territory, the grave assessment of casualties—all of these made for exciting television. And the excitement registered in ratings. The CBS special "48 Hours on Crack Street," for example, was the highest rated of any similar program in five years.[34] News agencies' interest in the dramatic and sensational also helps to explain why journalists were loathe to drop their dubious claims concerning the epidemic nature of crack use, the random violence it spawned, and the inevitable harm caused to infants exposed to cocaine in utero.[35]

By 1987–1988, however, some stories on the drug crisis were more reflective and nuanced. A few reporters began to point out, for example, that the crack "epidemic" was in reality concentrated in a handful of inner-city neighborhoods.[36] Some also began calling attention to the opportunistic uses of the "cocaine crisis" by politicians and the mass media. While many reporters who engaged in such second thoughts demonstrated no penchant for irony—they uniformly failed to recognize their own complicity in promoting drug hysteria—they did introduce the notion that crack is as much a "political" issue as a public health or criminal justice problem. Overall, though, the news media generally reinforced the notion that the best solution to the crime and drug problems is enhanced punishment and control.

Media Coverage and the Political Process

Media coverage can influence policy in a number of ways. First, media coverage may influence policy makers directly, independent of any impact on public opinion. In accounting for their legislative initiatives on crime and drugs, for example, politicians in the 1980s often cited increased media coverage of the drug problem as evidence of public concern to which they claimed to be responding. Officials may also perceive a high degree of media interest as an opportunity for political exposure or as a sign that public concern is likely to increase in the future. Thus, independent of its potential effects on public opinion, media coverage may influence the policy-making process at both the local and federal levels.[37] Second, while it is too simple to say that media discourse *causes* changes in public opinion, it is undoubtedly a crucial component of the context in which political opinions are formed. It is quite likely that the media's reproduction of the official view of crime and drugs played an important role in generating support for crime and drug policies aimed at punishment rather than prevention. The following chapter describes the social and cultural milieu in which these constructions were disseminated and analyzes their resonance with salient cultural themes and sentiments.

6

Crime and Punishment in American Political Culture

Although the adoption of tough anticrime and antidrug policies cannot be understood as primarily a response to popular sentiment, it is true that some segments of the American public have been quite receptive to law and order rhetoric and proposals. This receptivity is not the inevitable consequence of an unchanging and monolithic political culture, but reflects the ability of the conservative discourse on crime to address social and personal troubles in a compelling manner.

American Beliefs about Crime and Punishment

According to conventional wisdom, Americans have little patience with liberal explanations of criminal behavior and wholeheartedly support tough responses to crime. While there is evidence that public opinion has shifted in this direction, popular attitudes regarding crime and punishment have historically been—and continue to be—more complex and ambiguous than this view allows. The belief that criminal offenders should be severely punished, for example, coexists with widespread support for policies aimed at rehabilitation.[1] Many Americans continue to attribute crime to environmental and social conditions, a position typically associated with support for prevention and rehabilitation rather than punishment.[2] In fact, when asked to choose between spending money on prisons and police, on the one hand, or education and job training on the other, two-thirds of those polled in the late 1980s chose the latter.[3] Various perspectives on crime and punishment thus coexist in American political culture, even after decades of conservative political initiative on these issues. Arguments that depict law and order politics and policies as a direct manifestation of public attitudes oversimplify and dehistoricize American beliefs about crime and punishment. As Cullen and his colleagues conclude, "the structure of public attitudes is complex and could have supported progressive responses to the crime problem."[4]

It does appear, however, that support for punitive anticrime policies has grown in recent years. The percentage of Americans expressing support for capital punishment increased from 45% in 1965 to 71% in 1988 while the percentage reporting that the courts are "too lenient" grew from 48% to 82%.[5] Explanations of crime that highlight individual rather than social factors have also become more popular.[6] Most recently, the view that the main purpose of prisons is incapacitation has become widespread.[7]

It is important to note, however, that these shifts are fairly superficial. For example, support for capital punishment diminishes considerably when people are given the option of life sentences without the possibilty of parole.[8] Furthermore, the trend toward greater public punitiveness did not precede the adoption and implementation of tough anticrime policies; officials have played a crucial role in in framing the crime and drug issues in ways that imply the need for them. This initiative does not, in and of itself, explain the public's shift to the right (superficial as it may be): as many students of public opinion and mass communication processes point out, public opinion is not inevitably responsive to these kinds of political initiatives.[9] The success of the conservative campaign for law and order reflects the fact that this discourse makes sense of and provides a "solution" for pressing social and personal problems in ways that are compatible with popular wisdom and cultural beliefs and values.

Themes and Resources in American Political Culture

Some issue frames have an advantage because their ideas, symbols, and language resonate with larger cultural themes.[10] This does not mean that culture operates in a deterministic manner, but rather that myths, ideas, and sentiments serve as resources that can be drawn upon by actors attempting to promote particular interpretations of social issues.[11]

Individualism in American Political Culture

The discourse of law and order is predicated on that cluster of values and beliefs called "individualism," an orientation most observers agree is central to American political culture.[12] One well-known analyst summarized this philosophy as follows: "Individual persons must assume responsibility for whatever they have done; individuals are held accountable for the actions they have performed. Whether required by environmental constraints or not, individuals are held accountable for their actions of choice."[13]

The conservative discourse on crime and drugs—what Scheingold calls "volitional criminology"[14]—clearly resonates with (and reinforces) this set of beliefs and values. The *Respect for Authority* package, for example, explicitly rejects the idea that people are in any way influenced by their environment—and actually identifies the belief that persons are so influenced (as well as the "permissiveness" that accompanies this view) as an important *cause* of crime. Similarly, *Zero Tolerance* and *Get the Traffickers* emphasize that the decision to sell or use drugs is just that—a decision, an individual choice, presumably independent of socioeconomic factors such as unemployment and poverty. By contrast, *Poverty Causes Crime* suggests that people are constrained by their social environment.[15] Thus, the resonance of the conservative discourse with the individualistic orientation of American political culture may help to account for its popularity.

Morality, Family, and Authority in American Political Culture

In his recent analysis of popular views on crime and punishment, Sasson found widespread support for the belief that the collapse of traditional authority structures (especially the family) has caused a breakdown in the "moral fabric" of society, and that the crime and drug problems are manifestations of this moral collapse. As Sasson points out, this emphasis on family disintegration is not entirely consistent with either liberal ("structural") or conservative ("volitional") interpretations of crime. While the "social breakdown" perspective does explain crime in environmental terms, it also highlights the importance of normative integration and traditional authority structures rather than socioeconomic factors.[16]

Despite this ambiguity, there are important ways in which the conservative law and order approach is resonant with aspects of what Sasson calls the "social breakdown" perspective. In both discourses, crime is understood primarily as *immoral* behavior. Adherents of "social breakdown" emphasize the failure of the family to inculcate a proper moral sensibility in its members, lament parents' diminished authority over their children, and criticize the state for undermining this authority. The conservative emphasis on the need to combat "permissiveness" accords with this sense that traditional authority structures are collapsing. The conservative critique of the culture of welfare—with its assumption that public assistance programs cause moral and familial disintegration—also resonates with public concern about the collapse of the family. It appears, then, that the conservative discourse on crime resonates with and gives expression to the sense that the family and other structures that support conventional values are in decline.

The Fear of Crime

Americans are most afraid of being the victim of violent crime. While this fear does not necessarily lead to punitiveness, there is reason to suspect that heightened anxiety about crime may—under some circumstances—lead some to support efforts to crack down on criminals.

Evidence for this comes mainly from research regarding the effects of exposure to media depictions of crime. These studies report that heavy consumers of television crime shows—which focus disproportionately on violent crime and tend to depict law enforcement and punishment as the only means of dealing with this problem—are more fearful of crime, more likely to overestimate its prevalence, and more likely to have a "retributive justice perspective" (including high levels of support for punitive policies and opposition to gun control).[17] While some people are more vulnerable to these effects and some types of media products are more likely to produce them, it appears that exposure to images and stories that imply the need for enhanced enforcement and punishment leads some to become more supportive of punitive anticrime policies.[18]

These findings suggest that the conservative focus on street crime and the heightened fear that discussions of violent crime seem to engender—especially in the context of the mobilization of the law and order perspective—may also help to account for growing popular punitiveness. Under these conditions, it is quite possible that fear of violent crime—like American individualism and concern about social and familial breakdown—enhances the acceptability of the conservative discourse on crime.[19] Scheingold explains the connection between fear and punitiveness as follows: to the extent that we are concerned about our potential victimization, the "myth of punishment" provides some reassurance that something fairly immediate and direct can be done about the source of our anxiety.[20] Although this social-psychological explanation is compelling, it is important to remember that the fear of crime is not always associated with punitiveness and the latter may be prevalent even among those who are not particularly fearful.

In sum, the individualistic orientation of American political culture, concern about the breakdown of the family, and the fear of violent crime help to explain the growing popularity of the conservative approach to crime. One noteworthy fact is still unaccounted for, however: support for the get-tough approach has grown most dramatically among racially and socially conservative voters. To explain this, a brief discussion of trends in racial attitudes since the 1950s is necessary.

Race, Crime, and Punishment

Civil rights activists' efforts to highlight the immorality of the southern system of racial segregation and black disenfranchisement were, by almost any standard, quite successful. The brutal nature of southern white resistance to this movement engendered a high degree of support outside the South for the civil rights cause. White support for the principle of racial equality has continued to increase since that time: survey data suggest a dramatic shift away from the acceptance of segregation and discrimination and a strengthening belief in racial equality.[21]

Several factors complicate this happy story, however. Although support for the *principle* of racial equality remains strong, whites became less supportive of the goals of the black movement after 1965. For example, the percentage of Americans reporting that the Kennedy and Johnson administrations were "pushing integration too fast" increased from 32% in 1962 to 52% in 1966. The view that the "government had already done enough for blacks" also became widespread during this period.[22] In response to findings such as these some conclude that white anxiety about the speed and extent of racial reform was manifesting itself in opposition to policies aimed at the implementation of that principle. In their analysis of shifts in racial attitudes, Schuman, Steeh, and Bobo found significant evidence for this "underlying racism thesis."[23] In particular, these researchers found that whites have failed to support policies aimed at eradicating deeply rooted discriminatory practices, thus complicating their alleged commitment to racial equality.

White opposition to policies aimed at redressing racial inequality (including busing and affirmative action) has been interpreted in a variety of ways. Some argue that white antipathy to such programs is best understood as a manifestation of group conflict over scarce resources,[24] whereas others suggest that it is an expression of American individualism.[25] Another perspective suggests that this sentiment represents a mixture of antiblack affect and an individualistic culture.[26] Whatever the case, white resentment of programs aimed at actualizing racial equality has clearly grown since the mid-1960s. This hostility is quite relevant to the receptivity of some white Americans to the campaign for law and order.

Opposition to Racial Reform and Punitive Attitudes Toward Crime

Beliefs regarding crime and punishment are highly correlated with race and racial attitudes. Beginning in the early 1970s, researchers found that those expressing the highest degree of concern about crime also tended to oppose racial reform. For example, one study reported that 42% of those who strongly disapproved of racial and social reform efforts identified crime as the nation's most important problem (compared with only 13% of those who strongly approved of reform efforts).[27] Others also found that racial attitudes were an important predictor of support for law and order rhetoric:[28] "those who support a hard line on law and order issues tend to be more racist and sexist, tend not to support equal rights for unpopular minorities . . . and they have a more negative view of welfare recipients."[29] Opposition to busing was also strongly associated with white support for punitive anticrime policies.[30] More recently, Cohn and Barkan found that white support for punitive policies—and in particular, capital punishment—is highly associated with prejudice against blacks.[31] In sum, the conservative call for law and order appears to be most popular among those opposed to racial and social reform and who score higher on measures of racial prejudice.

The fact that blacks and whites tend to have different attitudes toward crime and punishment also suggests the importance of race on these matters. Despite the fact that blacks are far more likely to be victims of crime, whites are, on average, more punitive than blacks.[32] Analyses of legal decisions—including determinations of guilt and sentencing practices—also indicate that whites tend to be more punitive, particularly when the accused is black or Latino and the victim is white.[33] Insofar as the majority of Americans believe that most criminals are black and most victims are white, this pattern is quite important.[34] Race is also an important predictor of "criminal justice ideology" more generally: blacks are more likely to view the criminal justice system as discriminatory, to have an unfavorable view of the police, and to explain crime in terms of social conditions than are whites.[35]

In recent years, some researchers have found a "convergence" between levels of support for punitive policies among blacks and whites. For example, one study found that blacks became more likely to report that the courts are not harsh enough between 1981 and 1985 and that race became a weaker predictor of attitudes toward punishment during this period (although blacks were still less punitive than whites).[36] More recently, researchers found that blacks were 9.6% less likely than whites to believe that the courts were not harsh enough, a relatively small (though statistically significant) difference. Interestingly, these analysts found that while racial prejudice is most strongly related to punitiveness among whites, fear of crime is most highly related to support for punitive policies among blacks.[37] Thus, although there is some evidence that fear of crime is contributing to growing levels of support for punitive policies among blacks, racist attitudes continue to be the main determinant of white punitiveness.

In sum, attitudes regarding crime and punishment are inextricably bound up with race and racial attitudes; opposition to racial and social reform is crucial in accounting for white support for law and order policies. Not coincidentally, these socially and racially conservative voters have been the main target of the GOP's southern strategy discussed in chapter 2. As a result of its popularity among those swing voters—sometimes known as "Reagan Democrats"—the conservative mobilization of the crime issue has been an important source of support for the Republican party.

Race, Crime, and the Emerging Party Alignment

Beginning in the late 1960s political analysts began to argue that "the social issue"—riots, street crime, marches, drugs, rising taxes, and welfarism—had replaced traditional "bread and butter" issues as the most salient for Americans.[38] This shift coincided with the racial polarization of the two main parties: prior to 1960, the Republican and Democratic parties were perceived as equally likely to promote racial change, while after 1964 the public consistently identified the Democratic party as more supportive of racial reform. As one political analyst put it, "The most striking change has occurred on racial issues. In 1956, there was no consensus on the parties' stand on the issues of school integration and fair employment. . . . By 1968, there was a startling reversal in this judgement."[39] While the Democrats had traditionally been the party of "plain people," the conservative emphasis on the social issue—with all its racial connotations—threatened to increase white working and middle class support for the Republican party.

The rise of racial attitudes as the primary determinant of partisan loyalty and the association between racial attitudes and beliefs about crime and punishment help to explain the utility of crime-related issues to the Republican party. According to an analysis of the 1968 election, the Democrats lost most heavily among those who advocated the use of all available force to quell urban unrest and among people who thought that civil rights leaders were pushing their cause too fast. The salience of concern about urban unrest in the North and civil rights in the South meant that "beliefs concerning urban unrest, race, and crime had an unmistakable relationship to the vote."[40] Although the electoral benefits of the crime issue do not appear to have been as large in 1972, those who identified crime as the nation's most important problem favored the Republicans by a ratio of more than two to one in this election. Furthermore, 86% of those who reported that crime was the nation's most important problem at this time and switched parties shifted their allegiance to the GOP.[41] Both parties have since competed for these swing voters and this competition has clearly fueled efforts to be perceived as the toughest party on crime.

Understanding the Growing Significance of Race

The mobilization of the black movement and the backlash it engendered help to explain the increased political significance of racial attitudes, but the particular resonance of the conservative crime discourse with the swing voters discussed above requires explanation. In-depth interviews with these former Democrats indicate that many of these voters switched their allegiance to the Republican party largely as a result of their perception that the Democrats have granted minorities "special privileges." The authors of one report conclude that the belief that some (i.e., minorities) are "getting something for nothing" was crucial:

> The race issue, affirmative action, the sense of subsistence programs from the government going to 'people other than myself' . . . is a very prevalent theme all around the country. . . . What you hear is their hostility to a giveaway agenda for minority groups. . . . [T]hese Democrats talk about themselves as 'the people who work.' . . . Crime and drugs . . . have added a real dimension to this sense and to black-white relations.[42]

How is the perception that minorities have been the recipients of a "giveaway" agenda related to support for tough anticrime and anti-drug policies? As we saw in the earlier chapters, conservatives have depicted crime (and welfare) as a choice made by people looking for the easy way out. Involvement in drug sales in particular is often depicted as a preference motivated by greed and the hope of avoiding "real work." According to this view, it is the "law-abiding citizen" who bears the (economic and security) costs of these choices. By attributing the very real economic plight of "taxpayers" and "working persons" to the behavior of the "underclass," conservatives diminish the likelihood that these grievances will give rise to policies aimed at redistributing opportunities and resources in a more egalitarian fashion. Racial imagery has been crucial to the success of these efforts: whereas animosity toward those who "don't work" appears to be most pronounced among those crucial swing voters, the perceptions that inform them are widespread among whites. A 1990 National Opinion Research Center poll, for example, found that 62% of whites believe that blacks are lazier than whites and 78% believe that blacks are more likely to prefer being on welfare to being self-supporting.[43]

It thus appears that conservatives have been quite successful in attributing the plight of "the average American" to "cheats," "thieves," and "freeloaders," and in exploiting this sentiment for political and electoral gain.[44] In his analysis of the 1980 election, Burnham found that white working and middle class resentment of the (black) poor was the key determinant of those individuals' decision to "swing" Republican. Voters most likely to abandon the Democratic party were men from "middle America"—white, blue-collar workers, with a family income of $15,000–$25,000 and a high school education.[45] In particular, those voters who considered themselves "financially worse off" were most likely to shift their allegiance to the GOP. Burnam concluded that swing votes constitute "a rebellion of those who are [relatively] successful against the burdens imposed on behalf of those who are not."[46] The conventional distinction between "economic" and "social" issues does not capture the extent to which these types of issues have become ideologically intertwined: given their association with the poor, social issues such as crime, immigration, and welfare symbolize the costs of the underclass to "hard-working Americans."

The effectiveness of these discursive constructions thus lies in their capacity to explain the declining social and economic position of working people."[47] In fact, real wages have declined for many working persons—black and white—over the past several decades, and the tax burden of many working-class Americans has grown significantly.[48] These economic changes and the ascendance of the conservative interpretation of them help to explain growing receptivity to the discourse of law and order, as well as opposition to means-tested public assistance programs and efforts to limit immigration.[49] Race has been a crucial resource in these efforts to construct social and economic problems in ways that generate support for such measures: popular images of the welfare cheat and the criminal both tap and reinforce racial stereotypes and animosity.

Public Opinion in the American Political System

The existence of a two-party system in the United States creates a situation in which both parties compete primarily for the allegiance of same block of "independent" voters (the Democrat and Republican parties can take the allegiance of solidly liberal or conservative voters for granted). In the post–civil rights context, both parties have scrambled to woo those alienated by the Democratic inclusion of blacks and other minority voters—among whom the discourse of law and order is particularly popular. This dynamic helps to explain the unwillingness of most Democratic officials to offer competing conceptions of the crime and drug issues or advocate alternative solutions to these problems. The resonance of the conservative discourse on crime and drugs with more general cultural themes and the absence of a powerful or organized opposition to these campaigns have further increased their electoral viability.[50] The following chapter shows that the relative success of the campaign for law and order has had significant consequences for state policy.

7

Institutionalizing Law and Order

Between 1980 and 1994, the incarcerated population grew by 300% (from 500,000 to 1.5 million).[1] Federal and state prisons now house over one million prisoners, up from approximately 250,000 in 1970. The growth of the prison population in the U.S. is to a large extent a consequence of the war on drugs: between 1979 and 1994, the percentage of state inmates convicted of nonviolent drug offenses increased from 6% to nearly 30%; among federal inmates that percentage jumped from 21% to 60%[2] (and the jail population is also likely to include nonviolent drug offenders). Keeping this many people behind bars is, of course, quite costly: it is estimated that federal, state, and local governments spent more than $30 billion on corrections in 1994, up from $4 billion in 1975.[3]

The composition of the prison population has also changed dramatically. In 1930, 22% of all of those admitted to prison were black; by 1992, this percentage had climbed to 51%.[4] Again, the war on drugs appears to be the primary culprit: 90% of those admitted to prison for drug offenses in recent years were black or Hispanic.[5] And although the majority of those arrested in the course of the antidrug campaign have been male, the number of black (non-Hispanic) women incarcerated for drug offenses in state prisons increased by 828% between 1986 and 1991 alone.[6]

Increased frankness regarding the custodial (rather than correctional) nature of prisons and jails has accompanied the growth and changing racial composition of the prison population. While this move away from "corrections" has many causes, the rapid growth of the prison population has clearly reinforced it. In 1990, only nine states were operating their prisons at or below capacity; and nationwide, prisons are overcrowded by an average of 30%. In this context, resources for education, vocational training, and recreation within prisons have declined in both absolute and relative terms.[7] And if the recent display of bipartisan support for denying inmates Pell grants to pursue higher education is any indication, this trend is likely to continue in the future.

The number of persons on probation and parole has grown almost as rapidly as the prison population. By 1994, over five million adults (2.7% of the population) were under some form of correctional supervision.[8] Blacks and other racial minorities are disproportionately likely to be involved in these programs; in 1995, one out of three black male youths was under some form of state supervision.[9] And just as the "correctional" rationale for incarceration has been weakened, so too has the orientation of parole and probation shifted from rehabilitation and reintegration to management and supervision.[10]

The dramatic nature of these changes problematizes the argument that "for all its attention to street crime, the political process tends to dilute rather than mobilize purposeful political energy."[11] While it may be true that some anticrime measures are more symbolic than "practical," the politicization of crime and drug use at the national level served as an important catalyst for the expansion and reorientation of the federal, state, and local crime control systems.

The Federalization of Crime Control

Although Senator Barry Goldwater's bid for election to the presidency in 1964 was unsuccessful, his emphasis on the crime issue had a significant impact on the national political agenda. Shortly after his election—and in response to the perceived resonance of the conservative discourse on crime—President Lyndon Johnson created the Presidential Commission on Law Enforcement and the Administration of Justice to investigate the causes of criminal behavior and to propose policies aimed at reducing it. Johnson also called on Congress to create a federal pilot program to assist local crime control agencies in these efforts. The resulting Law Enforcement Assistance Act created the first federal grant program "designed solely for the purpose of bolstering state and local crime reduction programs,"[12] which was administered by the Office of Law Enforcement Assistance (OLEA). This program did not entail large sums of money and was never intended to be a major source of financial support for local law enforcement.[13] Nonetheless, the creation of the OLEA did establish a precedent for the federal government's assumption of responsibility for state and local crime control. It also had some unintended but important consequences that will be discussed in greater detail later in the chapter: it mobilized and strengthened the law enforcement and criminal justice lobbies and stimulated significant technical research on ways of improving the efficiency of law enforcement and criminal justice administration.[14]

In 1967, two years after its creation, the president's crime commission published an exhaustive account of its findings and recommendations. The authors of this report emphasized that crime was deeply rooted in the structure of American society and would not be significantly reduced by simply enhancing the efficiency of law enforcement or the criminal justice system. At the same time, many of the report's recommendations did imply "that crime control would be mainly accomplished by improving the criminal justice system" and "call upon the Federal goverment to support all components of the criminal justice system at the state and local levels."[15] In an effort to implement the Commission's recommendations and preempt the conservative manipulation of the crime issue in the upcoming elections, Johnson introduced the Safe Streets and Crime Control Act. This proposed legislation called for the establishment of a categorical federal assistance program to local governments, just as the Great Society programs funneled federal funds directly to urban communities. For obvious reasons, Republicans in Congress opposed these grant-in-aid programs and proposed an alternative revenue sharing system for distributing federal anticrime funds.

The ensuing debate thus centered on how—not whether—to allocate federal funds for state and local crime control. The final legislation established the Law Enforcement Assistance Administration (LEAA) within the Department of Justice to administer grants to states (which were to receive 60% of the available funds) and local governments (allocated the remaining 40%) for the development of comprehensive criminal justice plans. In the following years, federal funds allocated to state and local governments for this purpose increased dramatically, from approximately $300 million in 1968 to $1.25 billion in 1974.[16] Although the Safe Streets Act is considered the first major block grant program, it did not give states unlimited discretion over the allocation of federal funds, and congressional debates over whether these were best spent on law enforcement or the administration of criminal justice were often intense. Many advocates of the latter option were supporters of "community-based" correctional programs and were thus loosely affiliated with left-liberal critics of the criminal justice system. These disputes notwithstanding, the Safe Streets Act and the creation of the LEAA clearly established a new—primarily financial—role for the federal government in the fight against crime.

Drugs and the Federalization of Crime Control in the 1980s

As discussed earlier, the U.S. federalist system allocates to the national government a quite minor role in the fight against conventional street crime. In creating the funding programs described above, however, the federal government began the process of altering this institutional framework. Because the Harrison Narcotics Act of 1917 assigned the federal government a significant degree of responsibility for the enforcement of narcotics laws, crusading against drugs provided another means by which the federal government could exercise leadership in this arena. Indeed, as Zimring and Hawkins point out, the resurgence and expansion of the war on drugs represent "the most significant experiment in enlarging the federal lead in . . . crime control in this century" and was, in this sense, an obvious replacement for the war on crime of the 1960s.[17] As a result of federal leadership in the antidrug campaign, state and local law enforcement agencies have also made fighting drugs a top priority and most states have followed the federal government's lead in adopting tougher sentencing statutes for drug offenders.

The first major piece of anticrime legislation passed in the 1980s—the Comprehensive Crime Control Act of 1984—modified existing crime and drug policy in a number of ways. Bail procedures were altered to allow for the preventative detention of offenders classified as "dangerous," the sanction of criminal forfeiture was imposed for all felony drug offenses, and prison sentences and fines for the sale or distribution of controlled substances were increased. Less than two years later, Congress passed the Anti-Drug Abuse Act of 1986. This legislation substantially increased the maximum sentences for drug trafficking offenses, imposed mandatory minimum sentences for some drug trafficking and possession violations, broadened the scope of both criminal and civil forfeiture statutes, provided for the deportation of any alien who, after entry to the United States, became addicted to illegal drugs or who has at any time before or after entry been convicted of a controlled substance violation,[18] and dramatically increased mandatory sentences for the possession of small amounts of crack cocaine.[19] Some have argued that this legislation was passed primarily to "reassure" the public that "something was being done about crime."[20] While this argument appropriately acknowledges the symbolic dimension of crime and drug policy, it does not account for the extraordinary measures taken by Congress to ensure that similar measures were adopted and enforced by states and localities.

Asset Forfeiture and the Enforcement of the Law

Legal scholars have long recognized that the significance of legislation depends to a great extent on the likelihood that it will be enforced. Monetary incentives are one of the most effective mechanisms for encouraging enforcement of the law. Although both civil and criminal asset forfeiture statutes have a long history in American legal history, their use has become widespread only recently. The original RICO statute (Organized Crime Control Act of 1970) and the Comprehensive Drug Abuse and Control Act of 1970—both passed in 1970—included civil forfeiture provisions, some of which were specifically linked to the drug trade.[21] Even after the passage of this legislation, however, forfeiture provisions were rarely invoked.[22] Between 1970 and 1979, less than $30 million in assets had been seized by the government—the same amount forfeited in 1985 alone.[23]

In 1977, law enforcement agencies—and the DEA in particular—identified asset forfeiture provisions as a potential source of revenue and began to lobby Congress to broaden the conditions under which civil forfeiture statutes might be invoked.[24] Unlike their criminal counterparts, civil forfeiture statutes required only that law enforcement agencies show "probable cause" that the property in question was related to a drug or other criminal offense. In order to retrieve seized property, however, the owner is required to present a "preponderance of evidence" indicating that it was not in fact connected to a criminal enterprise.

In response to the lobbying efforts of law enforcement, Congress significantly expanded the scope of both civil and criminal forfeiture provisions in 1978.[25] These modifications allowed for the forfeiture of all profits from drug trafficking and all assets purchased with these profits, as well as all the monies intended to be used in exchange for illegal drugs—an arrangement that allowed the government to seize property never actually involved in illegal activities.[26] These provisions were further expanded in the 1984 Comprehensive Crime Control Act which permitted federal agents to seize goods valued at up to $100,000 without a full-scale court proceeding, established revolving funds in both the Justice and Treasury Departments so that assets seized by federal agents would remain within those agencies, and expanded the government's authority to require forfeiture of profits and proceeds from organized crime and narcotics enterprises.[27] Most important, this legislation provided the police with a direct incentive to identify drug law violators: local law enforcement agencies were entitled to deposit up to 95% of the value of the profits and proceeds seized in their own discretionary funds.[28] The Anti–Drug Abuse Acts of 1986 and 1988 expanded federal assets forfeiture provisions even further.[29]

The Bush administration also encouraged the states to pass their own asset forfeiture laws. By 1990, 49 states and the District of Columbia had statutes that allowed for the forfeiture of assets believed to be connected with illegal drug activity.[30] While civil forfeiture statutes allow for the seizure of property without the conviction or even arrest of its owner, these statutes do provide law enforcement with an incentive to identify drug offenders and therefore tend to increase the number of people arrested and prosecuted for violating narcotics laws. Criminal assets forfeiture statutes provide an even stronger incentive to identify and prosecute drug offenders.[31]

The broadening of asset forfeiture provisions and their adoption by states has been extremely lucrative for law enforcement agencies. Asset forfeiture receipts increased from $27.2 million in 1985 to $874 million in 1992; the assets and goods seized between 1985 and 1990 alone were estimated to be worth $4 to $5 billion. By 1990, over 90% of the police and sheriff's departments serving a population of at least 50,000 received money or goods from a drug asset forfeiture program.[32] Congressional expansion of asset forfeiture provisions also appears to have had a significant impact on the propensity of both federal and local law enforcement agents to enforce laws prohibiting the use and sale of illicit drugs. Indeed, as critics of these statutes point out, the lucrative nature of drug law enforcement means that law enforcement agencies may prioritize the prosecution of drug law offenders over those who commit violent offenses.[33]

Federal Financial Support for State and Local Crime Control

Direct federal funding for law enforcement and criminal justice programs is also an important means by which the federal government can influence state and local anticrime policy. Although the LEAA was disbanded in 1980, the Comprehensive Crime Control Act of 1984 re-established a federal grant program for state anticrime efforts, now administered by the Office of Justice Programs. The Anti–Drug Abuse Acts of 1986 and 1988 also provided funds to encourage the enforcement of drug laws and underwrite the escalating costs of prison construction. The 1986 legislation, for example, increased the authorization for state and local law enforcement by over one billion dollars and allowed these funds to be used for nonfederal prison construction.[34] In 1988, Congress established the Drug Control and System Improvement Grant Program, which provided funds to states and units of local government for carrying out specific antidrug efforts. As a result of this program, federal assistance to local law enforcement increased from $150 million in 1985 to $440 million in 1990.[35]

The 1994 Crime Bill was even more generous: this legislation allocated over $8 billion to the states to build and operate prisons and nearly $10 billion to hire more police officers.[36] These federal monies were allocated with strings attached. For example, funds for state prison construction are tied to specific "truth-in-sentencing" policies adopted by the federal government.[37] To the extent that the federal goverment is able to induce states to adopt tougher and more rigid sentencing policies, states' fiscal pressures will be exacerbated in the future. Indeed, these federal funds will diminish in the years ahead, leaving states in an even stickier financial situation than the one that compelled them to accept federal funds in the first place.

Federal Sentencing Guidelines

In 1984, after over ten years of deliberations, Congress enacted the most sweeping and dramatic reform of federal sentencing procedures—the Sentencing Reform Act. This legislation was a victory for the conservative wing of the movement for determinant sentencing: its primary goals included the reduction of unwarranted disparity and the correction of what was considered to be a pattern of undue leniency.[38] The act also called for the establishment of an independent agency (the U.S. Sentencing Commission) to develop guidelines that would structure judicial discretion and decision-making with respect to sentencing.

Simultaneous to the development and implementation of the federal sentencing guidelines, Congress began to adopt a series of "mandatory minimum" sentencing statutes. Although mandatory sentencing laws existed prior to this legislation, the enactment of these provisions was rare and was not aimed at whole classes of offenders.[39] In 1986, however, the Anti–Drug Abuse Act called for mandatory minimum (and stiffer) penalties for those who sold drugs to persons under 21, who employed a person under the age of 18 in a drug offense, or who carried a weapon while violating drug laws. Another provision of this legislation tied mandatory minimum penalties for drug trafficking directly to the amount of drugs involved, thereby limiting the range of factors that judges may consider in determining the appropriate punishment. Most significant, this legislation required a mandatory minimum sentence of five (and up to twenty) years for the simple possession of five or more grams of crack cocaine and twenty years for any offender who engaged in a "continuing drug enterprise."

In an attempt to encourage states to adopt these harsh mandatory minimum sentencing laws, the Bureau of Justice Assistance was authorized to make grants to states that adopted penalties similar to those established under the Anti-Drug Abuse Act. Congress also created a commission to develop a uniform code of state laws aimed at securing the federal government's stated objective: a "Drug Free America" by 1995.[40] Later, the federal government made conformity to federal sentencing standards a requirement for the receipt of federal funding. The Anti-Drug Abuse Act of 1988, for example, set clear standards for recipients of federal funds including the adoption of certain user sanctions and drug testing procedures. The 1994 Crime Bill went even further: states' eligibility for federal funds was now predicated upon their adoption of truth-in-sentencing laws and a "binding sentencing guideline system."[41] By 1994, 30 states had adopted variations of the "three strikes and you're out" laws that were the centerpiece of the 1994 Crime Bill.[42]

Consequences of the Federalization of Crime Control

The federal government's use of these mechanisms (as well as political dynamics at the state and local levels) has led to a dramatic increase in the number of drug arrests and convictions at all levels of the criminal justice system. The number of adult arrests for drug law violations reported by state and local police, for example, increased from 468,056 in 1981 to 1,008,347 in 1990.[43] Drug offenders make up an increasing percentage of all state prisoners: while the number of defendants sentenced to prison increased by 53% between 1981 and 1987, the number of drug offenders sentenced increased by 134% during that period.[44] The adoption of mandatory minimum sentences for drug-related offenses has meant that the increased number of drug arrests is an important cause of escalating incarceration rate.[45] A similar trend occurred in the federal court system, where the percentage of inmates convicted of violating drug laws is now over 60%.

As Michael Tonry points out in *Malign Neglect*, these developments have had particularly adverse—and foreseeable—consequences for African-American youth.[46] The war on drugs has been waged primarily in minority communities, despite evidence that the use of illegal drugs is evenly distributed by race.[47] Although survey data suggest that 13% of all monthly drug users are black, 35% of those arrested for drug possession, 55% of those convicted of drug possession, and 74% of those sentenced to prison for drug possession are black. Over 90% of all of those actually admitted to prison for all drug offenses are black or Hispanic.[48] As the authors of a recent report on the criminal justice system point out, this pattern reflects the extent to which law enforcement agents and prosecutors have concentrated on fighting drugs in minority communities: if the drug war had been waged on college campuses, its consequences would have been quite different.[49]

The particular focus on crack cocaine also helps to explain the racially disparate consequences of the war on drugs. Although the practice of smoking cocaine was not uncommon in middle- and upper-class communities in the late 1970s, the dangers associated with smoking cocaine received unprecedented publicity after a less expensive form of smokeable cocaine—"crack"—spread into the inner city in 1985. It is crack offenders who have born the brunt of the movement to get tough on drug law violators: as mentioned earlier, the 1986 Anti–Drug Abuse Act created stiff mandatory penalties (5–20 years) for the simple possession of less than $100 worth of crack cocaine.[50] Thus, for federal sentencing purposes, each gram of crack is the equivalent of 100 grams of powdered cocaine, despite the fact that most street level crack dealers buy their powder cocaine from large-scale retailers.[51] Crack cases are also treated more severely at the state and local levels, even where federal sentencing statutes have not been adopted: crack offenders are more likely to be charged with more serious offenses upon arrest, held in pretrial detention, indicted if arrested for a felony, and sentenced to jail or prison if convicted than those involved with powder cocaine or other illegal substances.[52] The focus on crack offenders has had distinctly different consequences for blacks and whites: African Americans are much more likely to appear in court as crack offenders, whereas whites are more likely to be prosecuted and sentenced under powder cocaine statutes.[53]

The Penal-Industrial Complex

As the criminal justice system grows, the size, resources, and authority of the interest groups that benefit from its expansion are also augmented.[54] These beneficiaries—including law enforcement, correctional workers, and a growing number of private firms—constitute what has come to be known as the "penal-industrial complex" and are now mobilizing to ensure that the wars on crime and drugs continue.

Law Enforcement as Political Lobby

A variety of developments in the 1960s—including Goldwater's politicization of the crime issue, the passage of the Law Enforcement Assistance Act of 1965, and a series of Supreme Court decisions aimed at expanding and protecting defendants' rights—prompted the growth and organization of a quite vocal law enforcement lobby.[55] Ironically, some of the first statements made on behalf of the new police lobby expressed concern about plans to increase federal funding for local law enforcement. In 1965, for example, the International Association of Police Chiefs passed a resolution against "any attempted encroachment by Federal government into State or local governments into the law enforcement field."[56] Over time, however, federal infusions of funds to state and local law enforcement were received with greater enthusiasm and appreciation. In monetary terms, these infusions were quite significant. By 1974, the police received 54% of the nation's $15 billion criminal justice budget, eight times the amount they received ten years earlier.[57]

The law enforcement lobby endeavors to protect and improve the working conditions of its members. As discussed earlier, the broadening of the asset forfeiture provisions was largely a response to the lobbying efforts of law enforcement agencies at the federal and state levels. The police lobby uses a wide range of tactics—including attempts to heighten the public's fear of crime—to achieve its goals. In a suburb of Washington, D.C., for example, the Fraternal Order of Police (FOP) responded to proposed layoffs and pay cuts for police officers by hiring a public relations firm. This company ran television commercials highlighting rising crime rates and the negative impact of proposed budget cuts on the ability of police officers to protect citizens from dangerous criminals. The FOP spent more than $10,000 in one week on this campaign and was successful in inducing politicians to back off the proposed budget cuts.[58]

The law enforcement lobby also promotes those politicians deemed sufficiently tough on criminals at the local, state, and federal levels. Politicians often compete to win the endorsement of law enforcement organizations, as George Bush and Bill Clinton did during the election campaign of 1992. Even though Clinton supported greater restrictions on handgun ownership—a position also endorsed by most law enforcement organizations—the national FOP gave its endorsement to President Bush. At a ceremony commemorating this endorsement, the president of the FOP praised Bush as "a great friend to law enforcement" and congratulated him for "proposing to the Congress the toughest crime bill in our history."[59] More recently, the FOP applauded President Clinton's decision to sign legislation maintaining tough mandatory minimum sentences for simple possession of crack cocaine.[60]

Correctional Agencies as Political Lobby

The dramatic expansion of the penal system was not anticipated by many of those initially advocating that federal funds be administered to state and local criminal justice agencies. Indeed, the 1967 President's Commission on Law Enforcement recommended the adoption of diversionary programs with the hope that the use of these alternatives would reduce the size of the conventional correctional apparatus. Throughout the 1960s and 70s, reformers both within and outside the LEAA proposed that federal funds be used to divert offenders from the criminal justice system; between 1968 and 1978, the LEAA funded over 1,200 "community" programs at an estimated cost of $112 million.[61]

But the creation of these programs did not lead to a reduction in the size of the crime control system. Instead, community-based alternatives became supplements to traditional criminal justice programs, resulting in a significant widening of the criminal justice "net."[62] The reasons for this are complex and have been the subject of much discussion. Clearly, the fact that funds earmarked for diversionary programs were allocated to criminal justice agencies (whose empire would have been reduced if diversion had been realized) had something to do with the ultimate fate of these programs.[63] Contradictions within the ideological foundation of the destructuring project may also have undermined the diversionary effort.[64] Most important, however, were factors external to this project, namely, the ascendance of the politics of law and order.

Indeed, the conservative emphasis on retribution, deterrence, and incapacitation translated quite directly into the expansion of the federal and state crime control systems. As a result, the number of people employed by the justice system has increased sharply, from 600,000 in 1965 to over two million in 1993.[65] The membership of the American Correctional Association more than doubled between 1982 and 1988 alone, while its budget increased from less than $1 million in 1979 to over $7 million in 1990.[66] By 1992, the number of full-time correctional employees was more than 500,000, more than are employed by any Fortune 500 firm other than General Motors.[67] Like the law enforcement lobby, correctional workers' associations are active at the local, state and federal levels, attempting to ensure the continued expansion of their employers and supporting politicians perceived as friendly to the criminal justice system.

For example, after California built sixteen new prisons in a sixteen-year period, the California Correctional Peace Officers' Association (CCPOA) emerged as one of the state's strongest political lobbies. The CCPOA is California's second most generous political action committee (PAC) and spends large amounts of money promoting the "victims' movement" and supporting "friendly" legislation and candidates. For example, the CCPOA was one of the largest contributors to the three-strikes-and-you're-out initiative and to Governor Pete Wilson's reelection campaign. These and other lobbying efforts appear to have been quite effective: 38 of the last 44 CCPOA-sponsored proposed bills were enacted by the legislature. As one analyst concluded, "Perhaps one of the reasons why corrections has become such a sacred cow in the state budget is because the prison guards' PAC contributes so generously to legislators and the Governor."[68]

Legislators promoting prison expansion are also winning the support of those who live in rural areas. Many of these financially strapped communities are competing to secure contracts for the construction of prisons in their "backyards." Some even go so far as to purchase the required land themselves and "donate" it to the state with the condition that it be used for prison construction.[69] This eagerness is not difficult to understand: prisons represent some of the first large-scale, unionized employers many rural areas have ever seen. As one official of the Federal Bureau of Prisons said, "[T]hey started realizing that we were a recession-proof, environmentally clean, attractive, safe industry."[70] Despite evidence that the benefits of prison construction for economically strapped rural communities are not as great as anticipated, many such communities continue to lobby for their construction.

Privatization, Prisons, and Profit

The expansion of the penal apparatus—and of prisons in particular—also ensures a market for private vendors of a wide array of goods and services. These companies range from financial firms competing for the opportunity to underwrite prison construction to private companies providing consulting, personnel management, architecture and building design, drug detection, medical, transportation, security, fine collection, bounty hunting, and food services. Defense companies are also jumping in on the action, aggressively marketing law enforcement equipment and other crime control devices. Indeed, many of these firms have created special divisions to retool their defense technologies for use by law enforcement and prison officials. These efforts are being encouraged and supported by the government: the National Institute of Justice held a major conference last year on "Law Enforcement Technology in the 21st Century," with special panels on "the role of the defense industry, particularly for dual use and conversion," and "how to penetrate the law enforcement market."[71] It appears that these entrepreneurial activities are paying off handsomely: in its publication "Outlook 2000," the Bureau of Labor Statistics ranked security as one of the twenty fastest growing service industries, only slightly behind data processing and computer software.[72]

Although the vast majority of prisons are still publicly owned and managed,[73] private prisons are growing and the rate of growth in private facilities is approximately four times greater than that of state prisons.[74] Given the increasing size of the prison population—which the 1994 Crime Bill will, in all likelihood, exacerbate—states are resorting to privatization as a way of relieving prison overcrowding and in the hopes of reducing the cost of getting tough.[75] The concern, of course, is that this practice amounts to a supply-side economic policy that will ensure the continual growth of the prison population.[76] For example, Florida recently contracted with Wachenhut Corrections Corporation (WCC) to manage several of its prisons. Since WCC is paid on a per-day, per-prisoner basis, the state of Florida guaranteed that the prison will never be less than 90% full.[77] Private firms may be able to achieve the same results themselves: these firms often have a say in awarding good-time credits and a voice in parole proceedings.[78] There is also evidence that the American "punishment industry" has begun to market its wares and exert influence over criminal justice policy in other countries as well.[79]

Developments in the Field of Penology

Like the political activism of the interest groups described above, developments within the field of penology both reflect and reinforce the expansion and reorientation of the crime control apparatus. In particular, the renewed emphasis on retribution, deterrence, and incapacitation served to legitimate the movement to get tough, while "managerial" penology helps to make this project more feasible.

The Decline of Rehabilitation and the Return of Retribution

At the same time that conservative politicians were lambasting liberal theories of crime and those who "mollycoddle" criminals, prisoners and their left-liberal allies were also developing a fairly devastating critique of the rehabilitative paradigm. In the early twentieth century, the hope of rehabilitating offenders led reformers to adopt indeterminant sentencing statutes and assign parole boards with the responsibility of determining when inmates were sufficiently rehabilitated to warrant release. Progressive critics of this system argued that its discretionary nature created space for arbitrary and discriminatory decision-making and that its therapeutic discourse obscured the exercise of power inherent in the act of punishment.[80] Many also criticized the positivist assumption that the criminal (as opposed to the system that labels some behaviors as criminal) is the proper object of knowledge for those studying crime and its control. As Stanley Cohen suggests, this critique paved the way for an older form of correctionalism: "[W]e performed elegant pirouettes around notions of freedom and determinism, and in the meantime let the state get on with its business of blaming and prosecuting."[81]

Not all critics of the rehabilitative project ignored the policy-making process. Advocates of the "justice model" agreed that rehabilitation should not be linked to sentencing decisions and proposed instead that the principle of "just deserts" serve as the basis for punishment.[82] As Greenberg and Humphries point out, the idea that people should be punished according to their "desert" has a critical edge in a system that punishes most poor people quite harshly for their crimes but affords rich people significantly more lenience.[83] But for reasons that are still the subject of debate, this potential was not realized.[84] Instead, the justice model's critique of indeterminant sentencing and rehabilitation, its emphasis on formal equality, and its proposal that retributive principles serve as the basis for punishment coincided with the conservative emphasis on the need to punish the individual wrongdoer more severely and to do so without consideration of the question of causation.

Both conservatives and progressives, then, offered a strong and compelling critique of the rehabilitative project in the 1960s and 1970s; all of the former and some of the latter also embraced the principle of retribution. This retributive orientation was the basis of the legislative reforms that began in the 1970s and accelerated—with much prodding and support by the federal government—in the 1980s and 1990s. Deterrence and incapacitation (public safety) also served as important rationales for the adoption of more punitive sentencing statutes. Research sponsored by right-wing think tanks and advocacy groups—as well as the Reagan-Bush Justice Department—has played an important role in legitimating these goals and the incarceration boom with which they have been associated.[85]

The New Penology

Students of contemporary social control patterns have noted that this punitive shift has been accompanied by the emergence of a more technocratic approach to crime control. This approach has been called "managerial" or "administrative" criminology, the "check 'em out" approach, and, most simply, the "new penology."[86] Advocates of the new penology profess no ideological affiliation, but see themselves as planners and systems engineers seeking to implement crime control policies aimed at the efficient management (rather than elimination or reduction) of criminal behavior.

A few features of this approach are worth highlighting. First, in the discourse of the new penology, the language of probability and risk supersedes any interest in clinical diagnosis, social context, or even retributive judgment. These "risk assessments" are based not on knowledge of the individual case but on actuarial or probablistic calculations.[87] Classification systems that use demographic and familial characteristics to assess offenders' "risk profile" reflect this orientation. Second, the new penology emphasizes efficiency and seeks to achieve it through the use of observational techniques calibrated according to assessments of risk.[88] The idea is that by differentiating between low-, medium-, and high-risk offenders and by subjecting each group to the appropriate level of supervision, crime control costs may be reduced. The hope of developing this capacity has stimulated a great deal of research: "Probably the most pervasive application of social science research in reform of the criminal justice system over the past twenty years has been the widespread incorporation . . . of research based on prediction and classification."[89]

The emergence of this managerial orientation as a dominant theme in American penology can be traced at least in part to the implementation of the Crime Control and Safe Streets Act of 1968. Prior to this time, few states, regional bodies, or localities engaged in comprehensive planning in the area of criminal justice. The federal government's requirement that these bodies develop such plans in order to qualify for federal funds, however, led to the emergence of a new profession—

criminal justice planning—by the mid-1970s.[90] These criminal justice experts emphasized the importance of developing "systems improvement projects" aimed at enhancing the coordination and efficiency of the criminal justice system and were more likely to be trained in operations research and systems analysis than in social work or sociology. Interestingly, the internal focus of this profession mirrors the Johnson administration's increasing unwillingness to emphasize the social context and causes of crime.[91]

As the 1970s and 1980s progressed, the new penology's promise to achieve greater efficiencies within the criminal justice system became especially appealing to government bodies adopting and implementing the get-tough policies promoted by the federal government. Observational and surveillance techniques such as electronic surveillance are significantly cheaper than custodial institutions.[92] The organizational requirements and fiscal implications of law and order policies fueled the interest in such low-cost alternatives.[93] Similarly, the search for the "career criminal" and the corresponding emphasis on selective incapacitation are largely motivated by the hope of reducing the expense associated with large-scale incarceration. Indeed, the possibility of an effective, selective incapacitation strategy is so appealing to many government agencies that "the criminal career notion . . . dominates discussion of criminal justice policy and . . . controls the expenditure of federal research funds."[94] These and other contemporary crime control strategies represent a distinct break with criminology's traditional interest in identifying the causes of crime.[95]

In sum, criminal justice policy and research is increasingly dominated by what Cohen calls the "conservative-managerial alliance."[96] In the context of the campaign for law and order, penological research has become less interested in the (social) causes of crime and more focused on the development of efficient crime-control policies. While a number of factors have contributed to the emergence of this more managerial and "realistic" approach, the shift of the epistemological gaze away from the criminal offender and his or her social context and the fiscal implications of the campaign for law and order have been especially important in this regard. The implications of these ideological and political shifts are considered in the final chapter.

8

Reconceptualizing the Crime Problem

In May 1995, the Federal Sentencing Commission recommended that Congress abandon those provisions of the Anti–Drug Abuse Act that penalize crack offenders more severely than any other type of drug law violator. The commission further suggested that those found with small amounts of crack or powder cocaine should be placed on probation as long as they did not also engage in violence or violate gun laws. In October—shortly after the Million Man March in Washington, D.C., in which the Reverend Jesse Jackson and others denounced crack sentencing laws as "unfair," "racist," and "ungodly"[1]—Congress voted to ignore the Sentencing Commission's recommendations and uphold existing sentencing statutes. President Bill Clinton subsequently approved Congress's decision, arguing that punishing crack offenders more harshly is appropriate because crack is more likely to be associated with violence and therefore takes a greater toll on the communities in which it is used and distributed.

That Congress and the president ultimately decided to uphold the sentencing laws that have contributed so much to the growth of the prison population and the incarceration of so many minority youths is not, at this point, surprising. What was more unusual was the debate triggered by the Sentencing Commission's recommendations and the congressional decision to ignore them. Suddenly, the overextension of the criminal justice system, its devastating effects on black youths, and the socioeconomic conditions that shape drug abuse were included in public discussions of the crime problem. While most of the figures participating in these discussions were high-ranking legal and political authorities, some grass-roots and community spokespersons also contributed to the discussion. The sudden outbreak of unrest in several federal prisons in response to Congress's decision drew further attention to the problem of racial inequities in the criminal justice system. In short, discussions of the crime issue were—for a short time—more democratic, less monolithic, and more likely to include "minority" perspectives on issues pertaining to criminal justice.[2]

That this debate marked such a dramatic break with previous discussions of the crime issue reveals how nearly hegemonic the law and order perspective has become. A political perspective may be characterized as hegemonic when "it is understood not just as one possible project among many alternatives, but as the only possible social order. Hegemony therefore involves a radical break with previously dominant discourses, and, at the same time, . . . the promotion of the sense that there is no alternative to the hegemonic project."[3] Until the Sentencing Commission's recommendations, the discourses of rehabilitation and root causes and other alternative ways of framing the crime problem were largely (although not entirely)[4] excluded from national political discourse on the crime issue. The irony, of course, is that current criminal justice policy as well as the ideology that justifies it is typically depicted as a direct expression of popular values—that is, of democracy-at-work.[5] But the ascendance of the rhetoric and policies of law and order is not an expression of democracy in action. Rather, this ideological framework was a component of the conservative project of state reconstruction: the effort to replace social welfare with social control as the principle of state policy.

This project is well under way. Between 1976 and 1989, the percentage of state budgets allocated to education and welfare programs declined dramatically—the former by 12% and the latter by 41%.[6] Across the states, the average monthly welfare benefit shrank from $714 to $394 (in 1995 dollars) between 1979 and 1993.[7] It is clear that the recent welfare "reform" legislation will accelerate this trend in the future. Meanwhile, state and federal "correctional" expenditures grew by 95% and 114% (respectively) between 1976 and 1989[8] and continue to increase dramatically. As we have seen, this shift has had a dramatic impact on the rate of incarceration in the United States. Indeed, if prison construction and the incarceration rate continue to increase at the present rate, two out of three young black men and one of four young Hispanic men will be in prison by the year 2020.[9] While the fiscal implications of these projections make their actualization unlikely, they nonetheless illustrate the nature and scope of this transformation of state expenditure and policy.

In the discourse that legitimates this transformation, the criminality, addiction, and delinquency of the impoverished—as well as their dependence on public assistance—symbolize their immorality, dangerousness, and preference for the "easy way out." As Feeley and Simon suggest, aspects of contemporary penological theory and practice—especially its rejection of the rehabilitative project—are also predicated upon this more pessimistic image of the underclass.[10] The get-tough and managerial discourses on crime are thus two sides of a remarkably cynical coin: both are fundamentally uninterested in the social causes of criminality or in reintegrating offenders and assume instead that punishment, surveillance, and control are the best response to deviant behavior. These assumptions—and the image of the undeserving underclass upon which they rest—have been made explicit in recent efforts to further increase the size of the prison population. Arguing that "all that is left of the 'black community' in some pockets of urban America is deviant, delinquent, and criminal adults surrounded by severely abused and neglected children, virtually all of whom were born out of wedlock," for example, Princeton Professor John DiIulio has mounted a quite energetic and high-profile defense of the massive use of imprisonment as an effective "crime-restraint tool."[11] As critics point out, these suggestions—apparently quite popular on Capitol Hill—involve "the wholesale writing off of a large segment of the population as irredeemably evil."[12]

The fact that some segments of the public have been receptive to this iconography of the underclass and the policies it implies does not mean that this approach is inevitable or unchangeable. As we have seen, political elites have shaped—not just responded to—public perceptions and sentiment on these issues. Those who attribute recent political developments to a universal and preexisting public desire surely overlook the historically contingent nature of political beliefs and attitudes and the importance of political leadership in shaping their expression.

Furthermore, while it is often assumed that popular attitudes are overwhelmingly punitive, more careful and nuanced analyses stress their complex and equivocal nature—even in the midst of the campaign for law and order. As discussed in chapter 6, public opinion polls that use fixed (rather than open-ended) questions (such as "Do you think the courts are too lenient, too harsh, or just about right?" or "Do you support the death penalty: yes or no"?) document a clear punitive trend in American attitudes about crime and punishment. At some level, then, Americans have become more responsive to rhetoric and policies associated with the campaign for law and order. But when methodologies that allow people to develop and express the complexity of their beliefs are used, a different picture emerges.[13] Research using these alternative techniques suggests that the popular desire for punishment coexists somewhat uneasily with support for prevention and rehabilitation and the belief that social conditions (such as poverty, unemployment, and a disrupted family life) are criminogenic. These studies also show that people's growing propensity to report that the courts are too lenient stems primarily from their tendency to overestimate the seriousness of most crimes and to underestimate the kinds of punishments they currently receive. The more exposure people have to nonsensationalistic accounts of real criminal incidents (from court documents rather than media accounts), the less punitive they become. And the public's willingness to advocate incarceration for nonviolent offenders and the death penalty for those convicted of homicide depends to a large extent upon whether they are presented with alternatives that ensure public safety.[14]

Public beliefs about crime and punishment, then, do not provide clear and unambiguous support for current criminal justice policies. Nor do popular sentiments on related issues point unequivocally toward these policies. Economic pressures, anxiety about social change, and a pervasive sense of insecurity clearly engender a great deal of frustration, and the scapegoating of the underclass has been a relatively successful way of tapping and channeling these sentiments. But this way of apprehending these realities is just that—a particular interpretation, one that has had significant consequences for state policy.

There is reason to believe that alternative crime frames might enjoy support from "experts" and the public alike. The view that crime has social causes and that certain kinds of rehabilitative programs are an effective means of responding to crime, for example, enjoys a significant degree of academic[15] and popular support. The notion that "family life" is an important dimension of the crime problem is also widespread, but has served primarily as a resource for advocates of the "culture of welfare" explanation of "underclass" behavior. But this need not be the case: one potentially fruitful strategy for progressives would be to stress the ways in which structural forces such as unemployment, low wages, inadequate medical care, and limited access to child care diminish the capacity of parents to care for their young.[16] Highlighting the impact of high rates of incarceration on individuals, families, and communities might also be a way of channeling concern about "social breakdown" in more progressive directions. The creation of a richer and more meaningful public discourse that includes these and other underrepresented perspectives is a first step toward the true democratization of crime and drug policy. This debate is not a peripheral one, but involves the very central question of whether state and social policy should emphasize and seek to promote inclusion or exclusion, reintegration or stigmatization. Nothing less than the true meaning of democracy is at stake.

Notes

Chapter 1

1. For example, the percentage identifying drugs as the nation's most important problem jumped from 3% in early 1986 to 64% in September 1989 (*New York Times*/CBS New Poll, pp. 2–4). More recently, the percentage of poll respondents identifying crime as the nation's most important problem increased from under 10% in June 1993 to over 50% by the August 1994 (Braun and Pasternak, "A Nation With Peril On Its Mind").

2. The percentage of Americans supporting capital punishment increased from 45% in 1965 to 71% in 1988, while the percentage reporting that the courts are "too lenient" grew from 48% to 82% (Niemi, Mueller, and Smith, *Trends in Public Opinion*). Americans are now more likely to believe that possession of small amounts of marijuana should be treated as a criminal offense and to explain crime in terms of individual rather than social factors. See Cullen, Clark, and Wozniak, "Explaining the Get-Tough Movement"; Mayer, *The Changing American Mind*; Roberts, "Public Opinion, Crime and Criminal Justice." While much of this research suggests that the public has become more punitive, there is also reason to suspect that this transformation is more superficial than is commonly supposed. This argument will be developed in chapter 8.

3. Chambliss, "Policing the Ghetto Underclass," p. 184; Oliver, "In the Government's Ongoing War Against Crime, Money Has Been Little Object," p. 1.

4. Tonry, *Malign Neglect*.

5. I am indebted to Cullen, Clark, and Wozniak for this term and their critique of the view that "current criminal justice policies are a direct reflection of

the increasing salience of lawlessness for citizens and their subsequent plea that the state punish and cage the wicked" ("Support for the Get-Tough Movement," p. 16).

6. Stinchcombe et al., *Crime and Punishment in America.*

7. Cullen, Clark, and Wozniak, "Support for the Get-Tough Movement"; Doble, *Crime and Punishment*; Flanagan, "Change and Influence in Popular Criminology"; Roberts, "Public Opinion, Crime and Criminal Justice."

8. Sasson, *Crime Talk.*

9. As Bennett points out, this "state of consciousness fallacy" oversimplifies and decontextualizes public sentiment: it "ignores the impact of political conditions on political thinking" and therefore overlooks "the range of characteristics that the public can exhibit in any given political situation." In contrast, Bennett's "situational" view of public opinion sees public attitudes as fluid, heterogeneous, and contextually defined. Bennett, *Public Opinion in American Politics.*

10. Bourdieu, *In Other Words*, p. 21.

11. Hall, *The Hard Road to Renewal.*

12. Ferrell and Sanders, *Cultural Criminology*. See also Barak, *Media, Process and the Social Construction of Crime*; Kappeler, Blumberg, and Potter, *The Mythology of Crime and Criminal Justice*; Scheingold, *The Politics of Street Crime: Criminal Process and Cultural Obsession*; Surette, *Media, Crime and Criminal Justice.*

13. Garland, *Punishment and Modern Society*, p. 20.

14. Kitsuse and Spector, "Toward a Sociology of Social Problems."

15. Edelman, *Constructing the Political Spectacle*; Gamson, *Talking Politics*; Gamson and Lasch, "The Political Culture of Social Welfare Policy"; Gamson and Modigliani, "The Changing Culture of Affirmative Action"; Gusfield, "Moral Passage"; Hilgartner and Bosk, "The Rise and Fall of Social Problems."

16. Ferrell and Sanders, "Culture, Crime and Criminology," p. 3.

17. Edelman, *Constructing the Political Spectacle*, p. 21.

18. See Corrigan and Sayer, *The Great Arch*, p. 3.

19. Gamson and Wolfsfeld, "Movements and Media as Interacting Systems"; Gitlin, *The Whole World is Watching*; McAdam, *Political Process and the Development of Black Insurgency.*

20. Bennett, *Public Opinion in American Politics*, p. 58.

21. Sasson, *Crime Talk.*

22. David Garland,"Penal Modernism and Postmodernism." Michel Foucault demonstrated that the emergence of this approach was predicated upon the ascendance of certain epistemological suppositions: that subjects are "knowable" and that problematic behavior has causes that may be identified and remedied by experts. The rise of this epistemological foundation was, in turn, linked to a set of sociohistorical developments that both reflect and facilitate the modern effort to solve social problems through the transformation of individuals. See Foucault, *Discipline and Punish.*

23. Rothman, *The Discovery of the Asylum*; Zimring and Hawkins, *Incapacitation*, pp. 6–10.

24. Rothman, *The Discovery of the Asylum*; see also Cullen and Gilbert, *Reaffirming Rehabilitation.*

25. On the decline of the rehabilitative project, see F. Allen, *The Decline of the Rehabilitative Ideal*; Bayer, "Crime, Punishment and the Decline of Liberal Optimism"; Cohen, *Visions of Social Control*; Cullen and Gilbert, *Reaffirming Rehabilitation*; Garland, *Punishment and Modern Society*, Chapter 1; Martinson, "What Works?—Questions and Answers About Prison Reform"; Zimring and Hawkins, *Incapacitation*. While it appears that criminal justice practitioners and politicians have lost interest in the rehabilitative project, Zimring and Hawkins provide evidence indicating that social scientific research continues to focus on this topic.

26. The reemergence of biological criminology may be an exception to this generalization. Many proponents of biological theories of crime argue that treatment is possible for those identified as biologically prone to criminality. However, both the historical record and the recent adoption of legislation mandating the "chemical castration" of certain sex offenders suggest that this research will primarily serve to justify different kinds of punishment rather than an alternative to it.

27. Cohen, *Visions of Social Control* and "Social Control Talk: Telling Stories About Correctional Change"; Feeley and Simon, "The New Penology"; Gordon, *The Justice Juggernaut*; Simon, *Poor Discipline.*

28. Garland, *Punishment and Welfare*; see also Feeley and Simon, "The New Penology."

29. See, for example, Bayer, "Crime, Punishment and the Decline of Liberal Optimism"; Cohen, *Visions of Social Control.*

30. Garland, "Penal Modernism and Post-Modernism," p. 204.

31. Richard Nixon, quoted in Marion, *A History of Federal Crime Control Initiatives, 1960–1993*, p. 70.

32. I am indebted to Craig Reinarman for this succinct summary of the current political transformation. See Reinarman, "The Social Construction of an Alcohol Problem."

33. Melossi also argues that efforts to punish may be elites' response to perceived social crisis (see Melossi, "Gazette of Morality and Social Whip").

34. Other researchers have also linked the growth of the state social control apparatus to the attack on the welfare state. In the early 1980s, for example, a group of scholars argued that declining rates of profitability led many Western governments to adopt "reverse Keynesian"—i.e., monetarist—economic policies. This shift corresponded to an increased reliance upon the coercive mechanisms of the state, for several reasons: the politicization of crime and punishment masks the extent of economic crisis by deflecting attention away from economic issues, identifies "target groups" which may be held responsible for social and economic problems, and helps to legitimate the state's authority (Horton, "The Rise of the Right"; Platt and Takagi, "Law and Order in the 1980's"; Ratner and McMullan, "Social Control and the Rise of the Exceptional State in Britain, the United States and Canada"). While I understand the

politicization of crime and punishment in the U.S. to be a response to political developments—especially the mobilization of the civil and welfare rights movements—rather than "the crisis of profitability," I am nonetheless indebted to these scholars' contention that crime-related issues are inextricably bound up with larger debates over state policy. For a more recent version of this argument, see Box, *Recession, Crime and Punishment*; Brake and Hale, *Public Order and Private Lives*; Hale, "Economy, Punishment and Imprisonment.

35. As Winant suggests, the ability to represent social issues in racial terms is central to the current pattern of conservative hegemony. Winant, "Difference and Equality," p. 119. This argument is supported by a recent study which found that television viewers are significantly less likely to advocate "social" rather than "individualistic" solutions to poverty when the welfare recipients depicted are black. See Iyengar, *Is Anyone Responsible?* p. 60. Similarly, there is evidence that a relatively large black population is one of the strongest predictors of states' incareration rates (McGarrell, Institutional Theory and the Stability of a Conflict Model of the Incarceration Rate").

36. Katz, "The Urban 'Underclass' as Metaphor of Social Transformation," p. 4.

37. Although sometimes referred to as the "correctional- industrial" complex, this term better captures the true nature and function of the criminal justice system in the 1990s. In fact, some prisoners' organizations have successfully sued prisons that call themselves "correctional" facilities, arguing that the absence of rehabilitative programs renders this term inappropriate.

38. Feeley and Simon, "The New Penology."

Chapter 2

1. Raymond Michelowski, "The Contradictions of Corrections," in *Making Law*, p. 100. In this article Michelowski critiques his earlier acceptance of this argument, asserting that in doing so he erroneously "treated such things as the 'fear of crime', [and] the rise in popular demands for more punishment . . . as social facts, rather than ideologically rooted, and in some cases politically manipulated, ways of constructing a social reality" (p. 103).

2. Greenberg, *Crime and Capitalism*, p. 316.

3. Wilson, *Thinking About Crime*, p. xvi. For a compelling critique of this argument, see Chambliss and Sbarbaro, "Creating Moral Panic, Legitimizing Repression and Institutionalizing Racism Through Law."

4. For example, Zimring and Hawkins suggest that the federal government took action on drugs because "a high level of public concern produces government action in a political democracy," while Wisotsky suggests that the Reagan administration "harnessed a preexisting momentum for a crackdown on drugs." See Zimring and Hawkins, *The Search for a Rational Drug Control Policy*, p. xi; Wisotsky, "Crackdown," p. 890. Elsewhere, however, Zimring and Hawkins argue that neither the rate of crime nor popular concern about it can account for the increasing incarceration rate in the 1970s and 80s. Zimring and Hawkins, *The Scale of Imprisonment*.

5. For example, Niemi, Mueller, and Smith attribute increased fear of crime and support for punitive policies to "objective shifts in the level of criminal activity," as evidenced by UCR data *(Trends in Public Opinion*, p. 133). Given that national survey data indicate that rates of drug use were more generally declining throughout the 1980s, the war on drugs is more difficult to explain in these terms.

6. See especially Barak, *Media, Process and the Social Construction of Crime*; Edelman, *Constructing the Political Spectacle*; Ferrell and Sanders, *Cultural Criminology*; Fishman, "Crime Waves as Ideology"; Hall et al., *Policing the Crisis*; Kappeler, Blumberg, and Potter, *The Mythology of Crime and Criminal Justice*; Jensen, Gerber, and Babcock, "The New War on Drugs"; Reinarman and Levine, "Crack in Context"; Sasson, *Crime Talk*; Scheingold, *The Politics of Street Crime*; Surette, *Media, Crime and Criminal Justice*.

7. In the crime case, twenty-nine polls that asked respondents "what do you think is the most important problem facing the nation" were taken. Twenty-five such polls were taken in the drug case. These polls were taken at three- to five-month intervals. The periodization of these case studies was designed to capture the rise and fall of public concern around each issue; these cases could not be extended without including lengthy periods in which the dependent variable remained at or close to zero.

8. For a more methodologically detailed version of this analysis, see Beckett, "Setting the Public Agenda."

9. These data are clearly imperfect. As discussed earlier, however, proponents of the democracy-at-work thesis generally point to the UCR data on violent crime to support their contention that concern about crime is linked to the rate of crime. My analysis is aimed at assessing the internal consistency of this argument.

It is difficult to gauge the accuracy of the UCR data on violent crime. The FBI UCRs are based on the number of crimes known to the police and reported by them to the FBI; the public's increased willingness to report their victimization, as well as improved recording techniques, may account for at least some portion of the increase reported in the 1960s and 1970s (see O'Brien, *Crime and Victimization Data*; President's Commission on Law Enforcement and the Administration of Justice, *The Challenge of Crime in a Free Society*). More serious crimes are more likely to be reported to and recorded by the police; the rate of violent crime is therefore probably less influenced by changing reporting and recording practices. However, there is evidence that cultural changes have had a significant impact on the willingness of rape victims to report their victimization to the police (see Orcutt and Faison, "Sex Role Attitude Change and Reporting of Rape Victimization, 1978–1985"). Homicide data are generally seen as more reliable because of the seriousness of the crime and the difficulties involved in concealing corpses.

10. Information regarding the incidence of drug use was taken from the National Institute on Drug Abuse (NIDA) survey ("The Household Survey on Drug Abuse"). The percentage of survey respondents aged twelve and over reporting illicit drug use in the past month was used as an indicator of the rate

of drug use. While DAWN data report the number of drug-related emergency room visits and therefore better capture the intensity of the drug problem, these data were not collected regularly throughout the period analyzed. In addition, methodological changes in NIDA's estimation procedures mean that the DAWN data collected before and after 1990 are not comparable. These data will, however, be considered in the discussion.

11. Information regarding the outcome variable—public concern—was taken from the Gallup Poll and the *New York Times*/CBS News Polls. Both of these are national public opinion surveys in which respondents are asked the open-ended question, "[W]hat do you think is the most important problem facing the nation?" The percentage of respondents identifying "crime," "juvenile delinquency," "the breakdown of law and order," or "general unrest" (in the crime case) and "drugs" or "drug use" (in the drug case) as the nation's most important problem served as the measure of public concern.

12. The number of speeches, statements, policy initiatives, or summaries pertaining to crime or drugs made by federal officials and reported in the mass media was utilized as an indicator of "political initiative." Because I analyzed newspaper stories indexed under "crime in the U.S." and national television news broadcasts, fewer than 2% of the stories analyzed relied on local or state (as opposed to national) officials. These stories were eliminated from the analysis.

13. The number of stories indexed under "crime in the U.S." in the *New York Times Index* and under "drug abuse" and "drug trafficking" in the *Vanderbilt Television News Index and Abstracts* (weeknights only) served as an indicator of media coverage. (Because the *Vanderbilt Television News Index and Abstracts* began in 1968, it was not possible to analyze television news coverage of the crime issue). Only those stories in which officials did not serve as sources were included in this category. "Political initiative" and "media coverage" thus refer to two different kinds of media coverage.

14. Public concern is typically measured in terms of the percentage of poll respondents who identify an issue as the nation's most important problem (see Stinchcombe et al., *Crime and Punishment in America;* see also Chambliss and Sbarbaro, "Creating Moral Panic, Legitimizing Repression and Institutionalizing Racism Through Law"). The sharpness of the fluctuations in levels of concern about crime and drugs reflects, in part, the use of this measure: the percentage of poll respondents who identify a given problem as the nation's most important may shift suddenly as other issues assume prominence in political debate.

15. I was primarily interested in identifying those factors associated with short-term fluctuations in public concern; the regression results are therefore based on an analysis of the differenced data. Differencing is a technique used to eliminate the linear trend from the data. With the linear trend removed, the regression coefficients estimate the association between short-term fluctuations in the explanatory and outcome variables.

16. The single equation models used here assume a one-way causal relationship between the explanatory and outcome variables. This relationship,

however, is probably interactive. These models thus tend to overestimate the effects of the explanatory variables; the results should be interpreted with this upward bias in mind.

17. Because the outcome variable (the percentage of poll respondents identifying crime or drugs as the nation's most important problem) is lower bound at zero, the regression coefficients may be interpreted in the following manner. The coefficient for the crime rate in Column 1 is –.0077: for every unit increase in the crime rate, the odds that a person would identify crime as the nation's most important problem would decrease $e^{-.0077}$ or 1.007 times. A unit increase in media and political initiative, would, according to the regression coefficients, lead respondents to be 3.49 and 3.94 (respectively) times more likely to identify crime as the nation's most important problem. The coefficients in Columns 2 and 3 are very close to those in Column 1 and can therefore be interpreted similarly.

18. According to the results presented in Column 1, a single unit increase in the rate of drug use increased the odds that a respondent would identify drugs as the nation's most important problem by 1.015 times. A unit increase in media and political initiative increased the odds that a person would respond in this manner by 1.29 and 6.43 times.

19. Treaster, "Emergency Room Cocaine Cases Rise"; Treaster, "Emergency Hospital Visits Rise Among Drug Abusers."

20. U.S. Department of Justice, *The Sourcebook of Criminal Justice Statistics*, 1988; National Institute on Drug Abuse, "National Household Survey on Drug Abuse," 1988.

21. The existence of simultaneity bias between political initiative and public concern in the regression model indicates that the relationship between these variables is interactive.

22. Statistical techniques designed to estimate the effects of reciprocal causal relationships make an elaborate set of assumptions about the data and therefore introduce significant specification errors. Furthermore, the statistical properties of the techniques that propose to estimate reciprocal causal effects in small samples are unknown. Given these difficulties, a case study method will be used to explore the relationship between shifts in the level of public opinion and political initiative over time.

23. Fewer than 1% of the Gallup Poll respondents identified crime-related problems as the nation's most important in 1963 and 1964. See Stinchcombe et al., *Crime and Punishment in America*, p. 25.

24. Braun and Pasternak, "A Nation With Peril On Its Mind."

25. Moore, "Public Wants Crime Bill," p. 11. See also Alderman, "Leading the Public."

26. Browning and Cao, "The Impact of Race on Criminal Justice Ideology"; Hagan and Albonetti, "Race, Class and the Perception of Criminal Injustice in America."

27. Cohn and Halteman, "Punitive Attitudes Toward Criminals"; see also Secret and Johnson, "Racial Differences in Attitudes Toward Crime Control."

28. Taylor, Scheppele, and Stinchcombe, "Salience of Crime and Support for Harsher Criminal Sanctions."

29. Cullen, Clark, and Wozniak, "Explaining the Get-Tough Movement"; Cullen et al., "Attribution, Salience and Attitudes Toward Criminal Sanctioning"; Stinchcombe et al., *Crime and Punishment in America.*

30. In fact, Stinchcombe et al. suggest that low levels of fear, high levels of punitiveness, and opposition to gun control in rural areas are best understood as a "defense of a rural hunting culture" rather than a response to the crime problem (*Crime and Punishment in America*, p. 13).

31. Although most researchers have concluded that fear of crime and punitiveness are largely unrelated, there is evidence that under some circumstances, fear of crime may lead some to embrace punitive policies (Langworthy and Whitehead, "Liberalism and Fear as Explanations of Punitiveness"). Cohn and Halteman, for example, argue that increasing punitiveness among blacks is related to growing rates of black victimization ("Punitive Attitudes Toward Criminals"). Media researchers also point out that heavy television consumers are more fearful of crime and more supportive of tough criminal sanctions (see Surette, *Media, Crime and Criminal Justice*). As we have seen, however, this link is not inevitable but depends upon the way in which the threat of crime is understood (Cullen et al., "Attributions, Salience and Attitudes Toward Criminal Sanctions").

32. Reinarman, "The Social Construction of Drug Scares."

33. Goode, "The American Drug Panic of the 1980s."

Chapter 3

1. The methodology upon which this analysis is based is described in chapter 5.

2. Beale, "Federalizing Crime."

3. Fogelson, *Big City Police*; Walker, *A Critical History of Police Reform.*

4. Growing numbers of arrests may have been the product of broader definitions of crime and increased law enforcement efforts. See Walker, *A Critical History of Police Reform*, p. 152.

5. Douthit, "Police Professionalism and the War Against Crime in the US, 1920's-30's."

6. Fogleson, *Big City Police.*

7. Cipes, *The Crime War*; Fogelson, *Big City Police*; Woodiwiss, *Crime, Crusades and Corruption.*

8. While civil rights activists did break southern state laws, they did so to draw attention to the unconstitutionality of those laws. Thus, it is not clear that this set of tactics can be accurately described as "criminal" or even as "civil disobedience."

9. Cronin, Cronin, and Milakovich, *The U.S. Versus Crime in the Streets.*

10. Justice Charles Whitaker, quoted in "Blamed in Crime Rise," *U.S. News and World Report*, p. 15.

11. "Lawlessness in U.S.—Warning From a Top Jurist," *U.S. News and World Report*, p. 27.

12. Nixon, "If Mob Rule Takes Hold in the US," p. 64.

13. Quoted in Caplan, "Reflections on the Nationalization of Crime, 1964–8," p. 585.

14. "Goldwater's Acceptance Speech to GOP Convention," *New York Times*, July 17, 1964.

15. "Goldwater at Illinois State Fair," *Chicago Tribune*, August 20, 1964.

16. Quoted in Rosch, "Crime as an Issue in American Politics."

17. As Caplan concludes, "it is ironic that the man who championed the issue that was ultimately to result in the creation of a new federal agency with an annual budget of nearly 900 million dollars was a strict constructionalist with a long record of opposition to expanding federal power" ("Reflections of the Nationalization of Crime, 1964–8," p. 586).

18. Both the President's Commission on Crime and the FBI concluded that more accurate reporting to and by police officials accounted for the increasing rate of property crime. See President's Commission on Law Enforcement and Administration of Justice, *The Challenge of Crime in a Free Society*; Caplan, "Reflections of the Nationalization of Crime, 1964–8," p. 586.

19. Omi, *We Shall Overturn;* Omi and Winant, *Racial Formation in the United States*.

20. Ibid., p. 17.

21. Quoted in Ginsberg, "Race and the Media."

22. Katz, *The Undeserving Poor*, p. 5.

23. Ibid., pp. 185–6. Katz and others show that the focus on the alleged misbehaviors of the poor has been central to their reconstruction as an undeserving underclass. See Gans, *The War Against the Poor;* Morris, *Dangerous Classes;* Schram, *Words of Welfare*. Crime, drug use, and delinquency were among the most important of these misbehaviors.

24. Quoted in Edsall and Edsall, *Chain Reaction*, p. 51.

25. Quoted in Baker, *Miranda*, p. 245.

26. Piven and Cloward, *Poor People's Movements*.

27. Gans, *The War Against the Poor*, p. 6; see also Karst, *Law's Promise, Law's Expression*, pp. 139–40. As Simon points out, "the criminal is the essential rebel against the modern orthodoxy of work" (Simon, *Poor Discipline*, p. 40). Increasingly, women who receive public assistance are also chastised for their alleged "laziness"; hence, the popularity of "workfare programs" designed to get "America working again."

28. Quoted in Matusow, *The Unraveling of America*, p. 143.

29. Piven and Cloward, *Poor People's Movements*, p. 338.

30. Moynihan, *The Politics of a Guaranteed Income*, p. 42.

31. Hall, "Moving Right," p. 122. Hall argues that "authoritarian-populism" (and Thatcherism in particular) represents a "new discursive articulation between the liberal discourses of the 'free market' . . . and the conservative themes of tradition, family and nation, respectability, patriarchalism and order." See

also Hall, *The Hard Road to Renewal*, p. 17. The parallel between the British and American cases is quite evident.

32. Quoted in Caplan, "Reflections of the Nationalization of Crime, 1964–8," p. 587.

33. President's Commission on Law Enforcement and Administration of Justice, *The Challenge of Crime in a Free Society*.

34. Johnson, "Remarks on the City Hall Steps, Dayton, Ohio," p. 1371.

35. Bayer, "Crime, Punishment and the Decline of Liberal Optimism."

36. The accuracy of these reports is debatable. While a 1965 Gallup Poll found that for the first time in many years, some Americans identified crime as the nation's most important problem this percentage was small (under 5%). This percentage grew, however, after the liberal conversion: by summer 1968, 29% identified crime or lawlessness as the most important problem facing the nation and 63% of those polled felt that the courts' were too lenient (up from 48% in 1965). See Niemi, Mueller, and Smith, *Trends in Public Opinion*, p. 136.

37. Quoted in "President Forms Panel to Study Crime Problems," *New York Times*.

38. Johnson, "Special Message to the Congress on Law Enforcement and the Administration of Justice," p. 264.

39. Bayer, "Crime, Punishment and the Decline of Liberal Optimism," pp. 184–8.

40. Ibid., p. 186.

41. Quoted in Marion, *A History of Federal Crime Control Initiatives, 1960–1993*, p. 70.

42. Quoted in Matusow, *The Unraveling of America*, p. 401.

43. Republican National Party, "Republican Party Platform of 1968," p. 987.

44. Quoted in Finkenauer, "*Crime as National Political Issue, 1964–76*," p. 21.

45. See chapter 2, Figures 3 and 4.

46. Cronin, Cronin, and Milakovich, *The U.S. Versus Crime in the Streets*.

47. Ibid.

48. Epstein, *Agency of Fear*, p. 65.

49. Quoted in Cronin, Cronin, and Milakovich, *The U.S. Versus Crime in the Streets*.

50. Quoted in Epstein, *Agency of Fear*, p. 69. Under Nixon, Title II of the Omnibus Crime Bill of 1968, allowing for the use of confessions obtained "voluntarily" but without the use of Miranda warnings, was implemented. In addition, the application of the Fourth Amendment's prohibition of unreasonable searches and seizures was narrowed and bail procedures were modified to facilitate the "preventative detention" of "dangerous offenders." While these changes significantly diminished defendants' civil liberties, they did not have a demonstrable effect on the rate of arrest, conviction, or incarceration.

51. Quoted in Cronin, Cronin, and Milakovich, *The U.S. Versus Crime in the Streets*, p. 84.

52. Milakovich and Weis, "Politics and Measures of Success in the War on Crime." See also Wright, *The Great American Crime Myth*, pp. 35–7, for a dis-

cussion of other innovative techniques used by the Nixon administration to create the impression that the rate of crime was decreasing.

53. As Zimring and Hawkins point out, the traditional allocation of crime control responsibilities "is one in which the federal goverment plays a distant secondary role to that of the states and local governments." But because the Harrison Narcotics Act established federal responsibility for the enforcement of narcotics laws, the federal goverment has played an important role in drug control throughout the twentieth century. See Zimring and Hawkins, *The Search for a Rational Drug Control Policy*, pp. 160–1.

54. Heymann and Moore note that the regulation of alcohol and other drugs has played an important role in expanding the scope of federal criminal jurisdiction ("The Federal Role in Dealing with Violent Street Crime," p. 105).

55. See Epstein, *Agency of Fear.*

56. Phillips, *The Emerging Republican Majority*, p. 39.

57. Ibid.

58. Reider, "The Rise of the Silent Majority," p. 243.

59. Guillory, "David Duke in Southern Context," p. 4.

60. Edsall and Edsall, *Chain Reaction*, p. 41.

61. Buchanan, *The New Majority.*

62. Phillips, *The Emerging Republican Majority.*

63. Ehrlichmann, *Witness to Power*, p. 233.

64. Edsall and Edsall, *Chain Reaction*, p. 150.

65. Crawford, *Thunder on the Right*, p. 176.

66. Such rhetoric is aimed at those who support, or think they should, the principle of racial equality but resent any policy or program that they understand as giving "preferential treatment" to minorities. Omi, *We Shall Overturn*, p. 120.

67. The GOP's new coalition is not the result of a wholesale conversion of working-class voters: the Republican party still enjoys its strongest support from the affluent. See Edsall and Edsall, *Chain Reaction*, p. 21.

68. Hall, "Moving Right."

69. For an insightful analysis of the importance of race in contemporary American politics, see Hadjor, *Another America.*

Chapter 4

1. *New York Times*/CBS News Poll, August 1990, pp. 2–4.

2. For example, the percentage of Americans who felt that "testing workers in general" for drug use would be an unfair invasion of privacy declined from 44% in 1986 to 24% in 1989. Similarly, the percentage of Americans who felt that possession of small amounts of marijuana should be treated as a criminal offense increased from 43% in 1980 to 74% in 1988. Gallup, *The Gallup Poll*, 1990.

3. Becker, *Outsiders*; Levine and Reinarman, "What's Behind Jar Wars."

4. Helmer, *Drugs and Minority Oppression*; Musto, *The American Disease*; Reinarman, "The Social Construction of Drug Scares."

5. Ibid.

6. See Chaiken and Chaiken, "Drugs and Predatory Crime"; Fagan, "Intoxication and Aggression."

7. Brownstein et al., 1992; Goldstein 1989.

8. Ronald Reagan, quoted in Kirschten, "Jungle Warfare," p. 1774.

9. Kirschten, "Jungle Warfare," p. 1774.

10. U.S. Department of Justice, *Attorney General's Task Force on Violent Crime*, p. v.

11. Pear, "Reagan's Advisor's Giving Top Priority to Street Crime and Victim's Aid," p. A1.

12. Davis, "The Production of Crime Policies," p. 127.

13. Quoted in Moore, *The Campaign For President*, p. 193.

14. Republican National Party, "The Official Report of the Proceedings of the Republican National Convention."

15. Reagan, quoted in Gross, "Reagan's Criminal Anti-Crime Fix."

16. Reagan, "Remarks at the Conservative Political Action Conference Dinner," p. 252.

17. Bush, "Address to Students on Drug Abuse," p. 747.

18. Reagan, "Remarks at the Conservative Political Action Conference Dinner," p. 253.

19. Reagan, "Remarks at the Annual Convention of the Texas State Bar Association in San Antonio," p. 1013.

20. Reagan, "Remarks at the Annual Conference of the National Sheriff's Association in Hartford, Conn.," p. 886.

21. Reagan, "Remarks at a White House Ceremony Observing Crime Victims Week," p. 553.

22. Reagan, quoted in Kirschten, "Jungle Warfare," p. 1774.

23. Reagan, "Remarks at the Annual Convention of the Texas State Bar Association in San Antonio," p. 1013.

24. Reagan, "Radio Address to the Nation on Proposed Crime Legislation," p. 1176.

25. Reagan, "Remarks to Members of the National Fraternal Congress of America," p. 1263.

26. Hale, "Economy, Punishment and Imprisonment," p. 342.

27. Reagan, "Remarks to Members of the National Governors Association," p. 238.

28. Reagan, "Remarks at a Fundraising Dinner Honoring Former Representative John M. Ashbrook in Ashland, Ohio," p. 672.

29. Lee Atwater, quoted in Edsall and Edsall, *Chain Reaction*, p. 145.

30. Bourdieu's conception of "habitus" helps to explain how actors develop perceptions of the world that are consistent with their interests but not based on a consciousness of them. "Habitus" refers to actor's "sense of the game or feel for the game, an intentionality without intention which functions as the principle of strategies devoid of strategic design, without rational computation and without conscious positing of ends." Bourdieu thus emphasizes the connection between one's "position" and categories of perception while rejecting

the notion that conscious motivation is necessarily involved: "[T]he legitimation of the social order is not the product of a deliberately biased action of propaganda or symbolic imposition; it results from the fact that agents apply to the objective structures of the social world structures of perception that have emerged from these objective structures and tend therefore to see the world as self-evident" (Bourdieu, *In Other Words*, p. 135).

31. Flanagan, "Change and Influence in Popular Criminology."

32. Pear, "FBI Chief Foresees Little Change Under Reagan."

33. Webster, "The Fear of Crime," p. 284.

34. "FBI Director Weighs War on Drug Trafficking," p. A11.

35. Executive Office of the President, "Budget of the U.S. Government."

36. The difference between the 1981 and 1991 figures is even more dramatic. Department of Defense antidrug allocations increased from $33 to $1,042 million in this period; DEA antidrug spending grew from $86 to $1,026 million; and FBI antidrug allocations grew from $38 to $181 million. See U.S. Office of the National Drug Control Policy, *National Drug Control Strategy*.

37. Ibid.

38. Stockman, *The Triumph of Politics*, pp. 153–4.

39. Ibid., p. 154.

40. Morganthau and Miller, "Turf Wars in the Federal Bureacracy."

41. Zimring and Hawkins make a similar point when they suggest that the "legalistic" (as opposed to the public health or cost-benefit) approach to drug use has dominated drug policy in the United States. The principal concern of the legalistic approach is the threat illegal drugs represent to the established order and the political authority structure. See Zimring and Hawkins, *The Search for a Rational Drug Control Policy*.

42. Reagan, "Remarks at a White House Kickoff Ceremony for National Drug Abuse Education and Prevention Week," p. 1330.

43. Regan, "Remarks at a White House Briefing for Service Organization Representatives on Drug Abuse," p. 1023.

44. Bush, "Remarks at the Acres Homes War on Drugs Rally in Houston, Texas," p. 1664.

45. Wisotsky, "Crackdown," p. 890.

46. Gallup, *The Gallup Poll*, 1990; see also Roberts, "Public Opinion, Crime and Criminal Justice."

47. For an extended discussion of the "crack panic" of 1986, see Reinarman and Levine, "Crack in Context."

48. Stutman, *Dead on Delivery*, p. 148.

49. Ibid., p. 217.

50. Danielman and Reese, "Intermedia Influence and the Drug Issue."

51. Fishman, "Crime Waves as Ideology," p. 538.

52. Ibid., p. 535.

53. Danielman and Reese, "Intermedia Influence and the Drug Issue"; Fishman, *Manufacturing the News*; Gans, *Deciding What's News*; Sigal, *Reporters and Officials*.

54. Brownstein, "The Media and the Construction of Random Drug Violence"; Orcutt and Turner, "Shocking Numbers and Graphic Accounts"; Reeves and Campbell, *Cracked Coverage*; Reinarman and Levine, "Crack in Context." Many of these analysts point out that the drug-related violence was depicted as a product of the chemical properties of these substances rather than the illegal—and hence violent—nature of the system by which they are distributed.

55. Brownstein, "The Media and the Construction of Random Drug Violence."

56. Merriam, "National Media Coverage of Drug Issues, 1983–1987."

57. Reagan, "Remarks to Members of the National Fraternal Congress of America," p. 1263.

58. "Reagan: Drugs Are the No. 1 Problem," *Newsweek*, p. 18.

59. Speakes, "Press Briefing by Larry Speakes."

60. Gallup, *The Gallup Poll*, 1990; *New York Times*/CBS News Poll, 1990.

61. During the 1988 presidential campaign, the Bush team used the Willie Horton incident as a way of mobilizing outrage about crime and blaming "liberal Democrats" for it. Horton was convicted of murder in Massachussetts, where Michael Dukakis was (later) governor. Subsequently released on furlough, Horton allegedly kidnapped a couple and raped the woman. The Bush campaign used Horton's photograph in its television spots describing the case (and Dukakis's role in it) because, as one of the producers of the advertisement put it, the incident was "a wonderful mix of liberalism and a big black rapist." (Quoted in Karst, *Law's Promise, Law's Expression*, p. 73–4). For many, the Willie Horton incident has become a shorthand means of describing the conservative effort to achieve political gain by manipulating the crime/race issue.

62. George Bush, quoted in Edsall and Edsall, *Chain Reaction*, p. 225.

63. See Gans, *The War Against the Poor*, p. 40.

64. Michelowski, "Some Thoughts Regarding the Impact of Clinton's Election on Crime and Justice Policy," p. 6.

65. Democratic National Committee, "The 1992 Democratic Party Platform."

66. Bill Clinton, quoted in "Clinton Nurtures High Hopes . . . ," *National Journal*, pp. 2794–5.

67. Idelson, "Democrat's New Proposal Seeks Consensus By Compromise."

68. Mauer, "The Fragility of Criminal Justice Reform," p. 17.

69. *Los Angeles Times* Public Opinion Survey, "National Issues."

70. For example, a *Washington Post*–ABC News Poll found that 39% of those polled trusted the Democrats to handle the crime problem, while 32% had more faith in the Republicans. See Poveda, "Clinton, Crime and the Justice Department," p. 76.

71. Idelson, "Tough Anti-Crime Bill Faces Tougher Balancing Act." A small group of liberal Democrats in the Senate did propose an alternative package aimed at improving police training, abolishing mandatory sentencing statutes, and tightening gun restrictions. Members of the Congressional Black Caucus also criticized the proposed legislation, especially its rejection of the Racial Justice Act, which would have allowed defendants to use evidence of racial bias to challenge their death sentences. Neither of these efforts was ultimately successful.

72. Masci, "$30 Billion Anti-Crime Bill Heads to Clinton's Desk."
73. Idelson, "Block Grants Replace Prevention, Police Hiring in House Bill."
74. Quoted in Kramer, "From Sarajevo to Needle Park," p. 29.
75. Poveda, "Clinton, Crime and the Justice Department," p. 75.

Chapter 5

1. Surette, *Media, Crime and Criminal Justice.*
2. McCombs and Shaw, "The Agenda-Setting Function of the Mass Media," p. 177.
3. Bennett, *Public Opinion in American Politics*; Iyengar and Kinder, *News That Matters*; Iyengar, Peters, and Kinder, "Experimental Demonstrations of the 'Not-So-Minimal' Consequences of Television News Programs"; Leff, Protess, and Brooks, "Crusading Journalism."
4. Iyengar, *Is Anyone Responsible?*.
5. Surette, *Media, Crime and Criminal Justice*. George Gerbner coined the term "mean world-view" to describe the outlook of heavy consumers of crime-related media products. Gerbner, "Television Violence."
6. See Roberts and Edwards, "Contextual Effects in Judgements of Crimes, Criminals and the Purpose of Sentencing"; Roberts and Doob, "News Media Influence of Public Views on Sentencing."
7. Ginsberg, *The Captive Public*, p. 32.
8. Burnham, *Critical Elections and the Mainsprings of American Politics.*
9. Ibid.; see also Chambers, "Party Development and the American Mainstream."
10. Quoted in Wasburn, *Broadcast Propaganda*, p. xvii; see also Burnham, *Critical Elections and the Mainsprings of American Politics*; Ginsberg, *The Captive Public*; Schudson, *Discovering the News.*
11. Schudson, *Discovering the News*, p. 151.
12. Ibid., pp. 162–3.
13. Ibid., p. 140. See also Epstein, *News From Nowhere*; Gans, *Deciding What's News*; Tuchman, *Making News*; Whitney et al., "Source and Geographic Bias in Television News 1982–4."
14. Nimmo, *Newsgathering in Washington.*
15. Whitney et al., "Source and Geographic Bias in Television News 1982–4," p. 170.
16. Sigal, *Reporters and Officials*, p. 195.
17. Gamson and Wolfsfeld, "Movements and Media as Interacting Systems"; Gitlin, *The Whole World Is Watching*; McAdam, *Political Process and the Development of Black Insurgency, 1930–70*; Ryan, *Prime Time Activism.*
18. For a more complete analysis of the media's coverage of the drug issue, see Beckett, "Media Depictions of Drug Abuse."
19. See Gamson and Lasch, "The Political Culture of Social Welfare Policy," pp. 399–401.
20. Kennan, "Liberal Looks at Violence in the U.S. and Where it is Leading."
21. Hoover, "Violence," p. 46.

22. Ibid., p. 45.

23. Katzenbach, "The Civil Rights Act of 1964," p. 28.

24. Johnson, "Annual Message to the Congress on the State of the Union," p. 7.

25. Reagan, "Radio Address to the Nation on Proposed Crime Legislation," p. 226.

26. Reagan, "Remarks to Members of the National Fraternal Congress of America," p. 1264.

27. George Bush, quoted in Nelson, "Bush Tells of Plan To Combat Drugs," p. A11.

28. My selection of sampling periods was guided by two main considerations: I attempted to identify 30-day periods at relatively even intervals with at least five media items. For the crime case, I analyzed news items appearing in August 1965, June 1968, August 1970, and March 1973. For the drug case, the sampling periods were October 1982; July 15– August 15, 1986; September 1989; and July 1991.

29. The *Television News Index and Abstracts* series began in 1968; it was therefore not possible to analyze television coverage of crime in the period from 1964 to 1974.

30. Television contains simultaneous audio and visual components, which may or may not display the signature elements of the same package; I therefore analyzed these components of television coverage separately. However, because there were no meaningful differences in the frame content of the audio and visual components of the crime and drug issues, I later collapsed these components into the category "television news."

31. Because I analyzed national news stories, the number of local or state officials cited in these stories was quite small. These stories were omitted from the analysis.

32. Nonstate sponsors included journalists, academics and other "experts," medical practitioners, lawyers and legal authorities, neighborhood organizers, drug users, crime victims, and others.

33. These findings are similar to those described by Reeves and Campbell. These authors argue that by reproducing the Reagan administration's depiction of drug use as criminal pathology, the network news media legitimated and intensified the "war on drugs." See Reeves and Campbell, *Cracked Coverage*.

34. Reinarman and Levine, "Crack in Context."

35. On this, see Beckett and Sasson, "The Media and the Construction of the Drug Crisis in America"; and Beckett, "Managing Motherhood."

36. Reeves and Campbell, *Cracked Coverage*.

37. See Surette, *Media, Crime and Criminal Justice*, pp. 87–9 for a discussion of this literature.

Chapter 6

1. Doble, "Crime and Punishment"; Cullen et al., "Public Support for Correctional Treatment"; Cullen et al., "Explaining the Get Tough Movement";

Roberts, "Public Opinion, Crime and Criminal Justice." In fact, most (64%) of those polled believe that most or some *violent offenders* can be rehabilitated, while only 6% believe than none can be (U.S. Department of Justice, *Sourcebook of Criminal Justice Statistics*, 1994, p. 176). As we will see in the following chapter, the extent to which researchers "find" this support appears to reflect both the wording and methodology used.

2. Cullen et al. found that those who attribute crime to psychological and social conditions are more likely to favor rehabilitation while those who accept the "classical" view that crime is a free choice tend to be more punitive ("Attribution, Salience and Attitudes Toward Criminal Sanctioning").

3. A poll taken in 1994 also found that the majority thought that investing in prevention and rehabilitation would be more effective than in punishment and enforcement. See U.S. Department of Justice, *Sourcebook on Criminal Justice Statistics*, 1994, pp. 171–2.

4. Cullen et al., "Explaining the Get Tough Movement," p. 22; see also Roberts, "Public Opinion, Crime and Criminal Justice." Gaubatz similarly emphasizes the complex and contradictory nature of attitudes about crime and punishment even among those who basically support get-tough rhetoric and policies. See *Crime in the Public Mind.*

5. See Niemi, Mueller, and Smith, *Trends in Public Opinion*, pp. 136–9. While the belief that the courts are too lenient increased most dramatically in the 1960s and 1970s, support for the death penalty has continued to increase throughout the 1980s.

6. While in the 1950s and 1960s theories that attributed crime and deviance to inadequate home life were most popular, many began to emphasize the importance of structural and economic conditions in the late 1970s and early 1980s. By the mid- to late-1980s, individualistic explanations had become more popular. Flanagan, "Change and Influence in Popular Criminology.

7. In 1982, a Harris poll found that 32% of poll respondents felt that the main function of prisons was "societal protection," while 19% identified "punishment" and 42% identified "rehabilitation" as the main purpose of incarceration (see Cullen et al., "Public Support for Correctional Treatment"). By 1994, 61% felt prisons functioned primarily "to keep criminals out of society," while 22% identified "punishment" and 13% "rehabilitation" as their main raison d'être (U.S. Department of Justice, *Sourcebook of Criminal Justice Statistics*, 1994, p. 177).

8. Sandys and McGarrell, "Attitudes Toward Capital Punishment."

9. See Bennett, *Public Opinion in American Politics*; Gamson, *Talking Politics*; Graber, *Processing the News*; Neuman, Just, and Crigler, *Common Knowledge*.

10. Gamson and Modigliani, "The Changing Culture of Affirmative Action," p. 169.

11. For a similar kind of analysis of the controversy over tort litigation reform, see Hayden, "The Cultural Logic of a Political Crisis."

12. Bellah et al., *Habits of the Heart*; Carbaugh, *Talking American*; Gans, *Middle Class Individualism*.

13. Carbaugh, *Talking American*, p. 45.

14. Scheingold, *The Politics of Street Crime*, chapter 1.

15. Sasson, *Crime Talk*, p. 150.

16. Ibid.

17. George Gerbner coined the term mean-world view to describe how the heavy consumption of media products affects people's perceptions of the world ("Television Violence"; see also Surette, *Media, Crime and Criminal Justice*.

18. It should be noted that these studies are correlational; this correlation is possibly spurious.

19. While I am suggesting that the fear of violent crime may be a resource for those promoting the get-tough approach to crime, it is also the case that the fear of violent (i.e., street) crime reflects the cultural and political preoccupation with this particular type of crime.

20. Scheingold, *The Politics of Law and Order*.

21. Schuman, Steeh, and Bobo, *Racial Attitudes in America*.

22. Orfield, "Race and the Liberal Agenda," pp. 334–6.

23. Schuman, Steeh, and Bobo, *Racial Attitudes in America*.

24. Bobo, "White Opposition to Busing"; Wellman, *Portraits of White Racism*.

25. Lipset and Schneider, "The Bakke Case."

26. Sears, "Symbolic Racism."

27. Furstenberg, "Public Reaction to Crime in the Streets."

28. Barkan and Cohn, "Racial Prejudice and Support for the Death Penalty for Whites"; Bennett and Tuchfarber, "The Social Structural Sources of Cleavage on Law and Order Policies"; Cohn and Halteman, "Punitive Attitudes Toward Criminals"; Corbett, "Public Support for 'Law and Order'."

29. Corbett, "Public Support for 'Law and Order'," p. 337.

30. Stinchcombe et al., *Crime and Punishment in America*.

31. Barkan and Cohn, "Racial Prejudice and Support for the Death Penalty By Whites."

32. Bennett and Tuchfarber, "The Social Structural Sources of Cleavage on Law and Order Polcies"; Browning and Cao, "The Impact of Race on Criminal Justice Ideology."

33. See Johnson, "Black Innocence and the White Jury" for a review of this literature.

34. Graber, *Processing the News*.

35. Browning and Cao, "The Impact of Race on Criminal Justice Ideology," p. 686. See also Hagan and Albonetti, "Race, Class and the Perception of Criminal Injustice in America."

36. Secret and Johnson, "Racial Differences in Attitudes Toward Crime Control."

37. Cohn and Halteman, "Punitive Attitudes Toward Criminals."

38. See especially Scammon and Wattenberg, *The Real Majority*.

39. Pomper, "Toward A More Responsive Two-Party System," p. 932.

40. Boyd, "Popular Control of Public Policy," p. 434.

41. Elliott, *Issues and Elections*.

42. Quoted in Edsall and Edsall, *Chain Reaction*, p. 226.

43. Thornton and Whitman, "Whites' Myths About Blacks."

44. Greenberg, "Plain Speaking."

45. Burnham, "The 1980 Earthquake," p. 106.

46. Ibid., p. 115.

47. This argument is consistent with conception of racism as a "culturally acceptable belief that defends social advantages" rather than simple bigotry or prejudice. See Wellman, *Portraits of White Racism*; see also Bobo, "White Opposition to Busing." While the emphasis on the structural sources of racism is compelling, his view that racism constitutes a defense of white privilege may obscure the extent to which this struggle over the "crumbs" itself results from the unequal distribution of resources.

48. Edsall and Edsall, *Chain Reaction*, p. 168.

49. Ibid., p. 152.

50. Joachim Savelsberg has argued that the relative weakness of neocorporatist institutions means that politicians in the U.S. are especially vulnerable to (apparent) shifts in public attitudes (Savelsberg, "Knowledge, Domination, and Criminal Punishment").

Chapter 7

1. Donziger, *The Real War on Crime*, p. 15; see also Irwin and Austin, *It's About Time*.

2. Ibid.; Chambliss, "Policing the Ghetto Underclass," p. 181; Irwin and Austin, *It's About Time*, p. 25. It is surprising to many that the cases involving the sale and possession of modest amounts of drugs are the most typical type of drug case prosecuted in the federal criminal justice system. See Zimring and Hawkins, *The Search For a Rational Drug Control Policy*, p. 162.

3. Holmes, "The Boom In Jails Is Locking Up Lots of Loot."

4. Donziger, *The Real War on Crime*, p. 103.

5. Ibid.

6. Mauer and Huling, "Young Black Americans and the Criminal Justice System."

7. Irwin and Austin, *It's About Time*, p. 224.

8. Donziger, *The Real War on Crime*, p. 15.

9. In some areas, the percentage is even higher (around 40%). Mauer, "The Drug War's Unequal Justice," p. 11.

10. See Simon, *Poor Discipline*, for an excellent analysis of this tranformation and its implications.

11. Scheingold, *The Politics of Street Crime*, p. 1.

12. Advisory Committee on Intergovernmental Relations, *Safe Streets Reconsidered*, p. 10.

13. Ibid. Between 1965 and 1968, OLEA expended $20.6 million dollars on 356 separate projects.

14. Ibid.; see also Feeley and Sarat, *The Policy Dilemma*, pp. 36–8.

15. Ibid., p. 11.

16. Ibid.

17. Zimring and Hawkins, *The Search For a Rational Drug Control Policy*, p. 158; see also Tonry, *Malign Neglect*.

18. William Weld, "Handbook on the Anti-Drug Abuse Act of 1986."

19. Bailor, "A Practitioner's Guide to the Anti-Drug Abuse Act of 1988."

20. Stolz, "Congress and Criminal Justice Policy Making."

21. Jensen and Gerber, "The Civil Forfeiture of Assets and the War on Drugs."

22. McAnany, "Assets Forfeiture as Drug Control Strategy."

23. Jensen and Gerber, "The Civil Forfeiture of Assets and the War on Drugs," p. 5.

24. See Wisotsky, "Exposing the War on Cocaine."

25. Law enforcement has also had a significant impact on state forfeiture provisions. In California, for example, police groups successfully lobbied the state legislature to reduce the state's burden of evidence from "beyond a reasonable doubt" to "more likely than not," decrease the amount of drugs necessary to trigger seizure of a vehicle, expand the kinds of property that can be seized, and increase the amount of money police and prosecutors can keep from 0% to over 90% (Webb, "Did Cops Get a License to Steal?").

26. Jensen and Gerber, "The Civil Forfeiture of Assets and the War on Drugs," p. 5.

27. "Enactment of Crime Package Culmination of 11 Year Effort," *Weekly Congressional Quarterly*.

28. McAnany, "Assets Forfeiture as Drug Control Strategy."

29. Criminal forfeiture provisions were amended to provide for the forfeiture of substitute assets and to include a broader range of violations that the legislation mandated must result in forfeiture. In addition, this legislation directed the Department of Justice to give higher priority to the use of civil forfeiture proceedings in order to fight drug trafficking and expanded the scope of seizable assets to include all property involved in drug law violations rather than just the gross receipts of such transactions. See Bailor, "A Practitioner's Guide to the Anti-Drug Abuse Act of 1988."

30. Jensen and Gerber, "The Civil Forfeiture of Assets and the War on Drugs," p. 7.

31. While it is likely that the incentive created by the asset forfeiture statutes does lead law enforcement to identify and arrest greater numbers of suspected drug law violators, most civil statutes allow property to be seized whether or not a person is charged or even arrested for violating a drug crime. Indeed, one study found that 80% of those whose property was seized under federal law were not charged with a related crime ("Bring Sanity to Seizure Laws"). This pattern is controversial for many reasons (see Jensen and Gerber, "The Civil Forfeiture of Assets and the War on Drugs"). One of the main issues involves the right to due process and the presumption of innocence: most civil statutes allow for the seizure of property based only on police "suspicion," and those whose assets are seized must sue the government for return of their as-

sets—but typically have been stripped of the assets with which they might mount such a lawsuit. In addition, there are increasing reports of persons who have been convicted of drug crimes being allowed to make a "donation" in exchange for a prison release or even a "clean slate" (see "Justice for Sale in Cobb").

32. Jensen and Gerber, "The Civil Forfeiture of Assets and the War on Drugs"; McAnany, "Assets Forfeiture as Drug Control Strategy."

33. Rasmussen and Benson, *The Economic Anatomy of a Drug War.*

34. Committee on Foreign Affairs, "Compilation of Narcotics Laws, Treaties and Executive Documents."

35. U.S. Department of Justice, *Sourcebook of Criminal Justice Statistics*, 1989.

36. Wright, "Federal Crime Bill."

37. Reitz, "The Federal Role in Sentencing Law and Policy," p. 117.

38. U.S. Sentencing Commission, "Mandatory Minimum Penalties in the Federal Justice System."

39. Ibid.

40. Stolz, "Congress and Criminal Justice Policy Making."

41. Wright, "The Federal Crime Bill."

42. Platt, "The Politics of Law and Order," p. 3.

43. U.S. Department of Justice, "Fact Sheet: Drug Data Summary."

44. U.S. Department of Justice, *The Sourcebook of Criminal Justice Statistics*, 1994, p. 494.

45. This argument is supported primarily by research on the impact of mandatory minimums in the federal system. For example, Meierhoefer found that the average sentence for drug behaviors that now carry a five-year mandatory prison term increased by 43%, from 42 months in 1964 to 62 months in 1990. Federal mandatory minimums have had a particularly significant impact on the sentencing of drug offenders: in 1990, 46.8% of all federal defendants were drug offenders, but 91% of all mandatory minimum defendants were drug offenders. See Meierhoefer, "The General Effect of Mandatory Minimum Prison Terms"; U.S. Sentencing Commission, "Mandatory Minimum Penalties in the Federal Justice System."

46. Tonry, *Malign Neglect*; see also Tonry, "Racial Politics, Racial Disparities, and the War on Crime."

47. U.S. Department of Justice, *Sourcebook of Criminal Justice Statistics*, Tables 3.103, 3.104, 3.105. Reprinted in Tonry, *Malign Neglect*, p. 110.

48. Donziger, *The Real War on Crime*, p. 103, 115.

49. Ibid., p. 115.

50. Bailor, "A Practitioner's Guide to the Anti-Drug Abuse Act of 1988."

51. Ironically, because the process of converting powder into crack cocaine involves the addition of adulterants, crack cocaine is typically much less pure than powder cocaine. Nonetheless, according to the U.S. Sentencing Commission Guidelines, the purity of the substance is irrelevant for sentencing purposes.

52. Belenko, Fagan, and Chin, "Criminal Justice Responses to Crack."

53. It is important to note that these prosecutions only partially reflect use patterns: survey data suggest that about half of all crack users are white, but

94% of those charged in the federal courts under crack statutes in 1994 were black. Morley, "Crack in Black and White."

54. Christie, *Crime Control As Industry*.

55. Feeley and Sarat, *The Policy Dilemma*, pp. 34–6.

56. Advisory Committee on Intergovernmental Relations, *Safe Streets Reconsidered*, p. 10.

57. Platt, "The Politics of Law and Order," p. 7.

58. This incident is reported in Chambliss, "Policing the Ghetto Underclass," p. 191.

59. Marcus and Devroy, "Police Group Gives Bush Its Blessing." It should be noted that not all measures supported by the FOP and other law enforcement organizations involve getting tough. For example, the FOP and other police organizations recently urged House Republicans to back off proposals to deny education to illegal immigrants, arguing that "forcing young people out of school and onto the streets would have disastrous long-term effects on public safety." Schmitt, "Police Scorn Plan to Deny Schooling to Illegal Aliens."

60. "Fraternal Order of Police Praises Clinton for Anti-Drug Stand," U.S. Newswire, Inc.

61. Austin and Krisberg, "Wider, Stronger, and Different Nets."

62. Ibid.; Cohen, *Visions of Social Control*.

63. Austin and Krisberg, "Wider, Stronger and Different Nets."

64. Cohen, *Visions of Social Control*.

65. Oliver, "In the Government's Ongoing War Against Crime, Money Has Been Little Object," p. 1.

66. As of 1988, there were 21,000 members in the ACA. "ACA Annual Report"; see also McMillan, "Diversified Resources Spell Success at ACA."

67. Donziger, *The Real War on Crime*, p. 93.

68. Quoted in Schiraldi, "The Undue Influence of California's Prison Guards' Union"; Macallair, "Lock 'Em Up Legislation Means Prisons Gain Clout."

69. See Donziger, *The Real War on Crime*, pp. 94–5.

70. Quoted in Elvin, "Correctional Industrial Complex Expands in U.S."

71. Ibid.

72. Quinson and Cox, "Making Crime Pay."

73. Ironically, the recent trend toward privatization was triggered by the expansion of community-based correctional programs in the 1970s. See Sellers, *The History and Politics of Private Prisons*.

74. As of 1994, the private sector managed 88 prison facilities incarcerating about 50,000 inmates (Donziger, *The Real War on Crime*, p. 88; for slightly different estimates see Oliver, "In the Government's Ongoing War Against Crime, Money Has Been Little Object," p. 1).

75. As Dr. Crants, chief executive at Corrections Corporation of America, put it, "Every elected official running in 1994 will be confronted by two main issues: fiscal restraint and violent crime. . . . The combination of the two is causing politicians to think about further privatization of correction facilities." Quoted in Quinson and Cox, "Making Crime Pay," p. C5. Executives of pri-

vate firms are certainly touting privatization as a way of reducing the cost of incarceration (see "Prison Privatization Is Efficient and Accountable, Wackenhut Corrections Chairman Maintains"). Because most such arguments do not take into account the cost of creating and operating a bureaucratic apparatus to monitor private institutions, it is unclear whether the state actually saves money by contracting with private firms to build and manage prisons.

76. Christie, *Crime Control as Industry*.

77. Thomas, "Making Crime Pay."

78. Oliver, "In the Government's Ongoing War Against Crime, Money Has Been Little Object," p. 1. Another concern is that private firms will simply fail to provide the services the state contracts with them to provide. In Texas, for example, Wackenhut Corrections Corporation is under investigation for allegedly misusing $700,000 of state funds allocated for running drug treatment programs. Although all of this money was to be spent on these programs, auditors discovered that it was actually being used to acquire a rare book collection and antique furniture for the firm's managers. Cohen, "Private Prison Firm Hit By Fraud Inquiry."

79. Lilly and Deflem, "Profit and Penality."

80. Bayer, "Crime, Punishment and the Decline of Liberal Optimism."

81. Cohen, *Against Criminology*, p. 122.

82. See Irwin, *Prisons In Turmoil*; Morris, *The Future of Imprisonment*; von Hirsch, *Doing Justice*.

83. Greenberg and Humphries, "The Cooptation of Fixed Sentencing Reform." Indeed, it is conceivable that a sentencing system based on deserts and selective incapacitation can lead to a reduction in the size of the prison population.

84. Greenberg and Humphries suggest that the justice models' individualistic orientation (i.e., its silence on the question of causation) and the decline of the social movements that might have pushed it in a progressive direction help to account for its cooptation (Ibid.). More generally, Cohen argues that the left's critique of positivism left questions of morality aside and thus conceded this terrain to the right (Cohen, *Against Criminology*, p. 119).

85. These scholars have primarily sought to show that current rates of imprisonment are an appropriate and effective response to the crime problem. For a description of these studies, see Mauer, "The Fragility of Criminal Justice Reform," pp. 18–9. For a critique of the research upon which these arguments rest, see Zimring and Hawkins, "The New Mathematics of Imprisonment."

86. See, respectively, Cohen, *Visions of Social Control* and *Against Criminology*; Gordon, *The Justice Juggernaut*; Feeley and Simon, "The New Penology."

87. Cohen, *Visions of Social Control*; Feeley and Simon, "The New Penology"; Simon, *Poor Discipline*.

88. Ibid.; see also Gordon, *The Justice Juggernaut*.

89. Gottfredson and Tonry, *Prediction and Classification*, p. vii; see also Feeley and Simon, "The New Penology." However, there is some evidence that rehabilitation remains an important research topic (see Zimring and Hawkins, *Incapacitation*, p. 13). These contrasting findings may reflect disciplinary dif-

ferences: Zimring and Hawkins analyzed the content of sociological abstracts, while others seem to be more focused on the criminal justice and penological literature. In addition, Zimring and Hawkins treated "rehabilitation/recidivism" as one category; their finding that the number of articles indexed under this heading remains high does not necessarily disprove the new penology thesis. Feeley and Simon do not suggest that interest in recidivism has waned, but that the nature of this interest has changed and now reflects the concern with prediction and risk management rather than evaluating the success of rehabilitative programs ("The New Penology").

90. Advisory Committee on Intergovernmental Relations, *Safe Streets Reconsidered*, p. 2, 175.

91. James Q. Wilson is, of course, a leading representative of the movement to abandon the search for the root (i.e. social) causes of crime (see especially *Thinking About Crime*).

92. Ibid.

93. Gordon, *The Justice Juggernaut*.

94. Gottfredson and Hirschi, "The True Value of Lambda Would Appear to Be Zero."

95. As was noted earlier, one type of research focusing on the causes of crime has become more popular in recent years: studies examining the role of biological factors, especially genetics, have become more widespread since the 1970s (Duster, *Backdoor To Eugenics*).

96. Cohen, *Against Criminology*, p. 20.

Chapter 8

1. Morley, "Crack in Black and White."

2. For a more complete discussion of this development, see Beckett and Sasson, 1997.

3. Smith, *New Right Discourses on Race and Sexuality*; see also Omi, *We Shall Overturn*; Omi and Winant, *Racial Formation in the United States*.

4. Some recent events have called attention to various—often hidden—dimensions of the crime issue. The controversy around the police beating of Rodney King in Los Angeles, for example, drew national attention to the problem of police brutality for some time. The O. J. Simpson trial—particularly after the taped interviews with Mark Furhman were released—also precipitated debate over the importance of race in the criminal justice system. Finally, some conservatives have recently joined progressives in advocating drug legalization and, in so doing, have called attention to previously ignored aspects of the war on drugs.

5. For a recent invocation of the argument that "lock 'em up" policies express unambiguous public sentiments, see DiIulio, "White Lies About Black Crime."

6. Steve Gold, cited in Donziger, *The Real War on Crime*, p. 48. This figure does not include Medicaid expenditures, which have increased due to rising health care costs.

7. Chomsky, "Rollback III."

8. Donziger, *The Real War on Crime*, p. 52.

9. Ibid., p. 106.

10. Feeley and Simon, "The New Penology," p. 467.

11. DiIulio, "White Lies About Black Crime," pp. 17–18. DiIulio also proposes the immediate and final removal of abused and neglected children (including those whose mothers test positive for drugs during pregnancy), the termination of all parental rights in such cases, and the creation of state-funded, church-run, and race-segregated orphanages. See DiIulio, "White Lies About Black Crime"; "The Question of Black Crime"; "Rescue the Young From Barbarism." It will be interesting to see if the ascendance of policies based on these assumptions stimulates a reevaluation of the rehabilitative project. While some of the most vociferous criticisms of the "therapeutic state" were made by the political left, current trends may lead some to conclude that the discourse of rehabilitation—with all its potential for justifying intrusion and the obfuscation of the exercise of power—also created some ideological space for efforts to reintegrate offenders and for the promotion of sociological explanations of crime. For a compelling version of this argument, see Cullen and Gilbert, *Reaffirming Rehabilitation.*

12. Shapiro, 'How the War on Crime Imprisons America," p. 17. See also Bendavid, "A Bull in Crime's China Shop."

13. These methodologies include open-ended survey questions, in-depth interviews, and focus group discussions. See Doble, "Crime and Punishment"; Gaubatz, *Crime in the Public Mind*; Haney, "Death Penalty Opinion"; McCorkle, "Punish and Rehabilitate?"; Roberts, "Public Opinion, Crime and Criminal Justice"; Sandys and McGarrell, "Attitudes Toward Capital Punishment"; Sasson, *Crime Talk.*

14. See Roberts, "Public Opinion, Crime and Criminal Justice," for an excellent overview of this literature.

15. See, for example, Hagan, *Crime and Disrepute*, chapter 3.

16. See Sasson, *Crime Talk* (chapter 9), for an elaboration of this argument.

Bibliography

"ACA Annual Report," (editorial), *Corrections Today* 50 (October 1988), p. 24.

Advisory Committee on Intergovernmental Relations, *Safe Streets Reconsidered: the Block Grant Experience 1968–75* (Washington, D.C., 1977).

Alderman, Jeffrey D., "Leading the Public: The Media's Focus on Crime Shaped Sentiment," *The Public Perspective*, 513 (March/April 1994), pp. 26–7.

Allen, Francis, *The Decline of the Rehabilitative Ideal* (New Haven: Yale University Press, 1981).

Austin, James, and Barry Krisberg, "Wider, Stronger, and Different Nets: The Dialectics of Criminal Justice Reform," *Journal Research in Crime and Delinquency* 18, 1 (January 1981), pp. 165–196.

Bailor, Bernard S., editor, *A Practitioner's Guide to the Anti–Drug Abuse Act of 1988* (Washington, D.C.: American Bar Association, Section of Criminal Justice, 1989).

Baker, Liva, *Miranda: Crime, Law and Politics* (New York: Atheneum Books, 1983).

Barak, Gregg, editor, *Media, Process and the Social Construction of Crime* (New York: Garland Publishers, 1994).

Barkan, Steven E., and Steven F. Cohn, "Racial Prejudice and Support for the Death Penalty by Whites," *Journal of Research in Crime and Delinquency* 31, 2 (May 1994), pp. 202–209.

Bayer, Ronald, "Crime, Punishment and the Decline of Liberal Optimism," *Crime and Delinquency* 27, 2 (April 1981), pp. 169–190.

Beale, Sara Sun, "Federalizing Crime: Assessing the Impact on the Federal Courts," *The Annals of the American Academy of Political and Social Science* 543 (January 1996), pp. 39–51.

Becker, Howard, *Outsiders: Studies in the Sociology of Deviance* (New York: Free Press, 1963).

Beckett, Katherine, "Setting the Public Agenda: 'Street Crime' and Drug Use in Contemporary American Politics," *Social Problems* 41, 3 (August 1994), pp. 425–447.

———, "Media Depictions of Drug Abuse: The Impact of Official Sources," *Research in Political Sociology* 7 (1995), pp. 161–182.

———, "Managing Motherhood: The Civil Regulation of Prenatal Drug Users, forthcoming in *Studies in Law, Politics and Society*, 16 (1997).

Beckett, Katherine, and Theodore Sasson, "The Media and the Construction of the Drug Crisis in America." Forthcoming in *The War on Drugs*, edited by Eric Jensen (ACJA Annual; Cincinnati: Anderson, 1997).

Belenko, Steven, Jeffrey Fagan, and Ko-lin Chin, "Criminal Justice Responses to Crack," *Journal of Research in Crime and Delinquency* 28 (1991), pp. 55–74.

Bellah, Robert N., Richard Madsen, William M. Sullivan, Ann Swidler, and Steven M. Tipton, *Habits of the Heart: Individualism and Commitment in American Life* (Berkeley: University of California Press, 1985).

Bendavid, Naftali, "A Bull in Crime's China Shop," *Legal Times*, February 12, 1996, p. 1.

Bennett, Lance W., *Public Opinion in American Politics* (New York: Harcourt Brace Jovanovich, 1980).

Bennett, Stephan Earl, and Alfred J. Tuchfarber, "The Social Structural Sources of Cleavage on Law and Order Policies," *American Journal of Political Science* 19 (1975), pp. 419–438.

"Blamed in Crime Rise: Civil Rights Excesses," *U.S. News and World Report* 62, 9 (February 27, 1967), p. 15.

Bloomberg, Thomas G., and Stanley Cohen, editors, *Punishment and Social Control: Essays in Honor of Sheldon L. Messinger* (New York: Aldine de Gruyter, 1995).

Bobo, Lawrence, "White Opposition to Busing: Symbolic Racism or Realistic Group Conflict?" *Journal of Personality and Social Psychology* 45, 6 (1983), pp. 1196–1210.

Bourdieu, Pierre, *In Other Words: Essays Towards a Reflexive Sociology*, translated by Matthew Adamson (Stanford, Ca.: Stanford University Press, 1990).

Box, Steven, *Recession, Crime and Punishment* (London: Macmillan Education, 1987).

Boyd, Richard, "Popular Control of Public Policy," *American Political Science Review* 66 (1972), pp. 429–444.

Brake, Michael, and Chris Hale, *Public Order and Private Lives: The Politics of Law and Order* (New York: Routledge, 1992).

Braun, Stephen, and Judy Pasternak, "A Nation With Peril On Its Mind," *Los Angeles Times*, February 19, 1994, pp. A1, A16.

"Bring Sanity to Seizure Laws," (editorial), *Chicago Tribune*, September 1, 1991, p. C2.

Browning, Sandra Lee, and Ligun Cao, "The Impact of Race on Criminal Justice Ideology," *Justice Quarterly* 9, 4 (December 1992), pp. 685–699.

Brownstein, Harry H., "The Media and the Construction of Random Drug Violence," *Social Justice* 18, 4, (1991) pp. 85–103.

Brownstein, Henry H., Patrick J. Ryan, and Paul Goldstein, "Drug-Related Homicide in New york City: 1984 and 1988," *Crime and Delinquency* 38 (1992), pp. 457–476.

Buchanan, Patrick, *The New Majority: President Nixon at Mid-Passage* (Philadelphia: Girard Bank, 1973).

Burnham, Walter Dean, *Critical Elections and the Mainsprings of American Politics* (New York: Norton Books, 1970).

———, "The 1980 Earthquake: Realignment, Earthquake, or What?" In *The Hidden Election*, edited by Thomas Ferguson and Joel Rogers (New York: Pantheon Books, 1981).

Bush, George, "Address to Students on Drug Abuse," *Public Papers of the Presidents 1989*, volume 1 (Washington, D.C.: Government Printing Office), pp. 746–749.

Bush, George, "Remarks at the Acres Homes War on Drugs Rally in Houston, Texas," *Public Papers of the Presidents 1989*, volume 2 (Washington, D.C.: U.S. Government Printing Office), pp. 1662–1664.

Caplan, Gerald, "Reflections on the Nationalization of Crime, 1964–8," *Law and the Social Order* 3 (1973), pp. 583–638.

Carbaugh, Donal, *Talking American: Cultural Discourse on Donahue* (New Jersey: Ablex Publishing Corporation, 1989).

Chaiken, Jan, and Marcia Chaiken, "Drugs and Predatory Crime." In *Drugs and Crime*, edited by Michael Tonry and James Q. Wilson (Chicago: University of Chicago Press, 1990).

Chambers, William Nisbet, "Party Development and the American Mainstream." In *The American Party Systems: Stages of Political Development*, edited by William Nisbet Chambers and Walter Dean Burnham (London: Oxford University Press, 1967).

Chambliss, William, "Policing the Ghetto Underclass: the Politics of Law and Law Enforcement," *Social Problems* 41, 2 (May 1994), pp. 177–194.

Chambliss, William J., and Edward Sbarbaro, "Creating Moral Panic, Legitimizing Repression and Institutionalizing Racism Through Law," *Socio-Legal Bulletin* (1993), pp. 4–12.

Chambliss, William J., and Marjorie S. Zatz, *Making Law: The State, the Law and Structural Contradictions* (Bloomington: Indiana University Press, 1993).

Chomsky, Noam, "Rollback III: The Virtual Collapse of Civil Society," *Z Magazine*, April 1995, pp. 17–24.

Christie, Nils, *Crime Control As Industry* (New York: Routledge, 1994), 2nd edition.

Cipes, Robert, *The Crime War* (New York: The New American Library, 1967).

"Clinton Nurtures High Hopes . . . ," *National Journal*, November 20, 1993, pp. 2794–2795.

Cohen, Nick, "Private Prison Firm Hit By Fraud Inquiry," *The Independent*, September 17, 1995, p. 11.

Cohen, Stanley, "Social Control Talk: Telling Stories About Correctional

Change." In *The Power to Punish: Contemporary Penality and Social Analysis*, edited by David Garland and Peter Young (Atlantic Highlands, New Jersey: Humanities Press, 1983).

———, *Visions of Social Control* (Cambridge: Polity Press, 1985).

———, *Against Criminology* (New Brunswick, N.J.: Transaction Publishers, 1992).

Cohn, Steven Barkan, and William Halteman, "Punitive Attitudes Toward Criminals: Racial Consensus or Racial Conflict?" *Social Problems* 38 (1991), pp. 287–296.

Committee on Foreign Affairs, "Compilation of Narcotics Laws, Treaties and Executive Documents" (Washington, D.C.: Congressional Research Service, 1986).

Corbett, Michael, "Public Support for 'Law and Order': Interrelationships With System Affirmation and Attitudes Toward Minorities," *Criminology* 19 (1981), pp. 328–343.

Corrigan, Philip, and Derek Sayer, *The Great Arch: English State Formation as Cultural Revolution* (Oxford: Basil Blackwell, 1985).

Crawford, Alan, *Thunder on the Right: The New Right and the Politics of Resentment* (New York: Pantheon Books, 1980).

Cronin, Thomas E., Tania Z. Cronin, and Michael Milakovich, *The U.S. Versus Crime in the Streets* (Bloomington: Indiana University Press, 1981).

Cullen, Francis, Gregory A. Clark, John Cullen, and Richard A. Mathers, "Public Support for Punishing White-Collar Crime: Blaming the Victim Revisited?" *Journal of Criminal Justice* 11 (1983), pp. 481–493.

Cullen, Francis, Gregory A. Clark, John B. Cullen, and Richard A. Mathers, "Attribution, Salience and Attitudes toward Criminal Sanctioning," *Criminal Justice and Behavior* 12, 3 (Sept. 1985), pp. 305–331.

Cullen, Francis T., Gregory A. Clark, and John F. Wozniak, "Explaining the Get-Tough Movement: Can the Public Be Blamed?" *Federal Probation* 45, 2 (June 1985), pp. 16–24.

Cullen, Francis T., and Karen E. Gilbert, *Reaffirming Rehabilitation* (Cincinnati: University of Cincinnati Press, 1982).

Cullen, Francis T., Sandra Evans Skovron, Joseph E. Scott, and Velmer S. Burton Jr., "Public Support for Correctional Treatment," *Criminal Justice and Behavior* 17, 1 (March 1990), pp. 2–18.

Danielman, Lucig H., and Stephen D. Reese, "Intermedia Influence and the Drug Issue: Converging on Cocaine." In *Communication Campaigns About Drugs: Government, Media and the Public,* edited by Pamela Shoemaker (Hillsdale, New Jersey: Lawrence Erlbaum Associates, Publishers, 1989).

Davis, David, "The Production of Crime Policies," *Crime and Social Justice* 20 (1983), pp. 121–137.

Democratic National Committee, "The 1992 Democratic Party Platform" (Washington, D.C.: Democratic National Committee, 1992).

DiIulio, John Jr., "The Question of Black Crime," *The Public Interest* 117 (Fall 1994), pp. 3–32.

———, "White Lies About Black Crime," *The Public Interest* 118 (Winter 1995), pp. 30–44.

———, "Rescue the Young From Barbarism," *The American Enterprise*, May/June 1995, pp. 32–33.

Doble, John, *Crime and Punishment: The Public's View* (New York: The Public Agenda Foundation, 1987).

Donziger, Steven R., editor, *The Real War on Crime: The Report of the National Criminal Justice Commission* (New York: HarperPerennial, 1996).

Douthit, Nathan, "Police Professionalism and the War Against Crime in the US, 1920's–30's." In *Police Forces in History*, edited by George Mosse (Beverly Hills: Sage Publications, 1975).

Duster, Troy, *Backdoor To Eugenics* (New York: Routledge, 1990).

Edelman, Murray, *Constructing the Political Spectacle* (Chicago: Chicago University Press, 1988).

Edsall, Thomas Byrne, and Mary Edsall, *Chain Reaction: The Impact of Rights, Race and Taxes on American Politics* (New York: Norton and Co., 1991).

Elliott, Euel W., *Issues and Elections: Presidential Voting in Contemporary America —A Revisionist View* (San Francisco: Westview Press, 1989).

Elvin, Jan, "Correctional Industrial Complex Expands in U.S.," *National Prison Project Journal* 10, 1 (Winter 1994/5), p. 2.

"Enactment of Crime Package Culmination of 11 Year Effort," *Congressional Quarterly Weekly Report* 42 (1984), pp. 21–25.

Epstein, Edward, *News From Nowhere: Television and the News* (New York: Random House, 1973).

———, *Agency of Fear: Opiates and Political Power in America* (New York: Random House, 1977).

Erlichmann, John, *Witness to Power: The Nixon Years* (New York: Simon and Schuster, 1970).

Executive Office of the President, "Budget of the U.S. Government" (Washington, D.C.: Government Printing Office, 1990).

Fagan, Jeffrey, "Intoxication and Aggression." In *Drugs and Crime*, edited by Michael Tonry and James Q. Wilson (Chicago: University of Chicago Press, 1990).

"FBI Director Weighs War on Drug Trafficking," *New York Times*, February 26, 1981, p. A11.

Feeley, Malcolm M.. and Austin D. Sarat, *The Policy Dilemma: Federal Crime Policy and the Law Enforcement Assistance Administration* (Minneapolis: University of Minnesota Press, 1980).

Feeley, Malcolm, and Jonathan Simon, "The New Penology: Notes on the Emerging Strategy of Corrections and its Implications," *Criminology* 30, 4 (1992), pp. 449–474.

Ferrell, Jeff, and Clinton R. Sanders, editors, *Cultural Criminology* (Boston: Northeastern University Press), 1995.

———, "Culture, Crime and Criminology." In *Cultural Criminology* (Boston: Northeastern University Press), 1995.

Finkenauer, James O., "Crime as National Political Issue, 1964–76," *Crime and Delinquency* 24, 1 (January 1978), pp. 13–27.

Fishman, Mark, "Crime Waves as Ideology," *Social Problems* 25 (1978), pp. 531–543.

———, *Manufacturing the News* (Austin: University of Texas, 1980).

Flanagan, Timothy, "Change and Influence in Popular Criminology: Public Attributions of Crime Causation," *Journal of Criminal Justice* 15 (1987), pp. 231–243.

Fogelson, Robert, *Big City Police* (Cambridge, Mass.: Harvard University Press, 1977).

Foucault, Michel, *Discipline and Punish* (translated by Alan Sheridan. (New York: Vintage Books, 1979).

"Fraternal Order of Police Praises Clinton for Anti-Drug Stand," U.S. Newswire, Inc., October 31, 1995.

Furstenberg, Frank, "Public Reaction to Crime in the Streets," *The American Scholar* 40, 4 (Autumn 1971), pp. 601–610.

Gallup, George, editor, *The Gallup Poll* (Wilmington, Del.: Scholarly Resources, 1990).

Gamson, William A., *Talking Politics* (Boston: Cambridge University Press, 1992).

Gamson, William A., and Kathryn E. Lasch, "The Political Culture of Social Welfare Policy." In *Evaluating the Welfare State: Social and Political Perspectives*, edited by Shimon E. Spiro and Ephraim Yuchtman-Yaar, (New York: Academic Press, 1983).

Gamson, William A., and Andre Modigliani, "The Changing Culture of Affirmative Action," *Research in Political Sociology* 3 (1987), pp. 137–177.

Gamson, William A., and Gadi Wolfsfeld, "Movements and Media as Interacting Systems," *The Annals of the American Academy of Political and Social Science* 528 (1993), pp. 114–125.

Gans, Herbert, *Deciding What's News* (New York: Vintage Books, 1979).

———, *Middle Class Individualism* (New York: The Free Press, 1988).

———, *The War Against the Poor* (New York: Basic Books, 1995).

Garland, David, *Punishment and Modern Society: A Study in Social Theory* (Chicago: University of Chicago Press, 1990).

———, "Penal Modernism and Postmodernism." In *Punishment and Social Control: Essays in Honor of Sheldon L. Messinger*, edited by Thomas G. Bloomberg and Stanley Cohen (New York: Aldine de Gruyter, 1995).

Garland, David, and Peter Young, editors, *The Power to Punish: Contemporary Penality and Social Analysis* (Atlantic Highlands, New Jersey: Humanities Press, 1983).

Gaubatz, Kathlyn Taylor, *Crime in the Public Mind* (Ann Arbor: University of Michigan Press, 1995).

Gerbner, George, "Television Violence: The Art of Asking the Wrong Question," *Currents in Modern Thought* (July 1994), pp. 385–397.

Ginsberg, Benjamin, *The Captive Public: How Mass Opinion Promotes State Power* (New York: Basic Books, 1986).

Ginsberg, Carl, *Race and the Media: The Enduring Life of the Moynihan Report*, Institute for Media Analysis (Washington, D.C.), no. 3, 1996.

Gitlin, Todd, *The Whole World is Watching* (Berkeley: University of California Press, 1980).

Goldstein, Paul L., Henry H. Brownstein, Patrick J. Ryan, and Patricia A. Belluci, "Crack and Homicide in New York City: A Conceptually-based Event Analysis," *Contemporary Drug Problems* 16, 4 (1989), pp. 651–687.

Goldwater, Barry, "Goldwater's Acceptance Speech to GOP Convention," *New York Times*, July 17, 1964.

———, "Goldwater at Illinois State Fair," *Chicago Tribune*, August 20, 1964.

Goode, Erich, "The American Drug Panic of the 1980s: Social Construction or Objective Threat?" *Violence, Aggression and Terrorism* 3 (1989), pp. 327–348.

Gordon, Diana R., *The Justice Juggernaut: Fighting Street Crime, Controlling Citizens* (New Brunswick, N.J.: Rutgers University Press, 1990).

Gottfredson, Don M., and Michael Tonry, editors, *Prediction and Classification: Criminal Justice Decision-Making* (Chicago: University of Chicago Press, 1987).

Gottfredson, Michael, and Travis Hirschi, "The True Value of Lambda Would Appear to Be Zero: An Essay on Career Criminals, Criminal Careers, Selective Incapacitation, Cohort Studies, and Related Topics," *Criminology* 24, 2 (1986), p. 213.

Graber, Doris, *Processing the News: How People Tame the Information Tide* (New York: Longman, 1984).

Greenberg, David F., *Crime and Capitalism: Readings in Marxist Criminology* (Palo Alto, Ca.: Mayfield, 1983).

Greenberg, David F., and Drew Humphries, "The Cooptation of Fixed Sentencing Reform." In *Crime and Capitalism: Readings in Marxist Criminology*, edited by David Greenberg (New York: Mayfield Publishing Company, 1993), 2nd edition.

Greenberg, Stanley B., "Plain Speaking: Democrats State Their Minds," *Public Opinion* 9 (Summer 1986), pp. 44–50.

Gross, Bertram, "Reagan's Criminal Anti-Crime Fix." In *What Reagan Is Doing To Us*, edited by Alan Gartner, Colin Greer, and Frank Riessman (New York: Harper and Row, 1982).

Guillory, Ferrel, "David Duke in Southern Context." In *The Emergence of David Duke and the Politics of Race*, edited by Douglas Rose (Chapel Hill: University of North Carolina Press, 1992).

Gusfield, Joseph, "Moral Passage: The Symbolic Process in Public Designations of Deviance," *Social Problems* 15 (1967), pp. 175–188.

Hadjor, Kofi Buenor, *Another America: The Politics of Race and Blame* (Boston: South End Press, 1995).

Hagan, John, *Crime and Disrepute* (Thousand Oaks, Ca.: Pine Forge Press, 1994).

Hagan, John, and Celesta Albonetti, "Race, Class and the Perception of Criminal Injustice in America," *American Journal of Sociology* 88, 2 (1982), pp. 329–355.

Hale, Chris, "Economy, Punishment and Imprisonment," *Contemporary Crises* 13 (1989), p. 342.

Hall, Stuart, *The Hard Road to Renewal* (London: Verso Press, 1988).

———, "Moving Right," *Marxism Today* 25 (1979), pp. 113–137.

Hall, Stuart, Chas Critcher, Tony Jefferson, John Clarke, and Brian Roberts, *Policing the Crisis: Mugging, the State and Law and Order* (New York: Holmes and Meier Publishers, Inc., 1978).

Haney, Craig, "Death Penalty Opinion: Myth and Misconception," *Criminal Defense Practice Reporter* 1995, 1 (January 1995), pp. 1–7.

Hayden, Robert M., "The Cultural Logic of a Political Crisis: Common Sense, Hegemony and the Great American Liability Insurance Famine of 1986," *Studies in Law, Politics and Society*, 11 (1991), pp. 95–117.

Helmer, John, *Drugs and Minority Oppression* (New York: Seabury Press, 1975).

Heymann, Philip B., and Mark H. Moore, "The Federal Role in Dealing with Violent Street Crime: Principles, Questions and Cautions," *The Annals of the American Academy of Political and Social Science* 543 (January 1996), pp. 103–115.

Hilgartner, Stephen, and Charles L. Bosk, "The Rise and Fall of Social Problems: A Public Arenas Model," *American Journal of Sociology* 94, 1 (July 1988), pp. 53–78.

Holmes, Steven A., "The Boom In Jails Is Locking Up Lots of Loot," *The New York Times*, November 6, 1994.

Hoover, J. Edgar, "Violence," *Vital Speeches of the Day* 35 (1968), pp. 42–68.

Horton, John, "The Rise of the Right: A Global View," *Crime and Social Justice*, Summer 1981, pp. 7–17.

Idelson, Holly, "Democrats' New Proposal Seeks Consensus By Compromise," *Congressional Quarterly* 51, 33 (August 14, 1993), pp. 2228–2289.

———, "Tough Anti-Crime Bill Faces Tougher Balancing Act," *Congressional Quarterly Weekly Reporter*, January 29, 1994, pp. 171–173.

———, "Block Grants Replace Prevention, Police Hiring in House Bill," *Congressional Quarterly*, February 18, 1995, pp. 530–532.

Irwin, John, *Prisons In Turmoil* (Boston: Little Brown, 1980).

Irwin, John, and James Austin, *It's About Time: America's Imprisonment Binge* (Belmont, Ca.: Wadsworth, Inc., 1994).

Iyengar, Shanto, *Is Anyone Responsible?* (Chicago: Chicago University Press, 1991).

Iyengar, Shanto, and Donald Kinder, *News That Matters* (Chicago: University of Chicago Press, 1987).

Iyengar, Shanto, Mark D. Peters, and Donald Kinder, "Experimental Demonstrations of the 'Not-So-Minimal' Consequences of Television News Programs," *American Political Science Review* 76 (1982), pp. 848–858.

Jensen, Eric L., and Jurg Gerber, "The Civil Forfeiture of Assets and the War on Drugs: Expanding Criminal Sanctions While Reducing Due Process Protections," *Crime and Delinquency*, July 1996 (forthcoming).

Jensen, Eric L., Jurg Gerber, and Ginna M. Babcock, "The New War on Drugs:

Grass Roots Movement or Political Construction?" *Journal of Drug Issues* 21 (1991), pp. 641–657.

Johnson, Lyndon B., "Remarks on the City Hall Steps, Dayton, Ohio," *Public Papers of the Presidents 1964*, volume 2 (Washington, D.C.: U.S. Government Printing Office, 1965), pp. 1368–1373.

———, "Special Message to the Congress on Law Enforcement and the Administration of Justice," *Public Papers of the Presidents 1965*, volume 1 (Washington, D.C.: U.S. Government Printing Office, 1966), pp. 263–271.

Johnson, Sheri L., "Black Innocence and the White Jury," *Michigan Law Review* 83 (1985), pp. 1611–1689.

"Justice for Sale in Cobb," (editorial), *The Atlanta Constitution*, April 9, 1994, p. A18.

Kappeler, Victor K., Mark Blumberg, and Gary W. Potter, *The Mythology of Crime and Criminal Justice* (Prospect Heights, Ill.: Waveland Press Inc., 1994).

Karst, Kenneth L., *Law's Promise, Law's Expression: Visions of Power in the Politics of Race, Gender and Religion* (New Haven, Conn.: Yale University Press, 1993).

Katz, Michael B., *The Undeserving Poor: From the War on Poverty to the War on Welfare* (New York: Pantheon Books, 1989).

———, "The Urban 'Underclass' as Metaphor of Social Transformation." In *The Underclass Debate*, edited by Michael B. Katz (Princeton: Princeton University Press, 1993).

Katzenbach, Nicholas, "The Civil Rights Act of 1964," *Vital Speeches of the Day* 32 (1964), pp. 27–29.

Keith, Michael, and Malcolm Cross, editors, *Racism, the City, and the State* (New York: Routledge, 1993).

Kennan, George Frost, "Liberal Looks at Violence in the U.S. and Where it is Leading," *U.S. News and World Report* 64 (June 17, 1968), pp. 66–69.

Kirschten, Dick, "Jungle Warfare," *National Journal*, October 3, 1981, p. 1774.

Kitsuse, John, and Malcolm Spector, "Toward a Sociology of Social Problems: Social Conditions, Value-Judgements, and Social Problems," *Social Problems* 20 (1973), pp. 407–419.

Kramer, Michael, "From Sarajevo to Needle Park," *Time Magazine*, February 21, 1994, p. 29.

Langworthy, Robert H., and John T. Whitehead, "Liberalism and Fear as Explanations of Punitiveness," *Criminology* 24, 3 (1986), pp. 575–592.

"Lawlessness in U.S.—Warning From a Top Jurist," *U.S. News and World Report*, September 6, 1965, p. 27.

Leff, Donna R., David L. Protess, and Stephen C. Brooks, "Crusading Journalism: Changing Public Attitudes and Policy-Making Agendas," *Public Opinion Quarterly* 50 (1986), pp. 300–315.

Levine, Harry, and Craig Reinarman, "What's Behind Jar Wars," *The Nation* 244, 12 (1987), pp. 388–90.

Lilly, J. Robert, and Mathieu Deflem, "Profit and Penality: An Analysis of the Corrections-Commercial Complex," *Crime and Delinquency* 42, 1 (January 1996), pp. 3–20.

Lipset, Seymour, and W. Schneider, "The Bakke Case: How Would the Case Be Decided At the Bar of Public Opinion?" *Public Opinion* 1 (1978), pp. 38–44.

Los Angeles Times Public Opinion Survey, "National Issues," January, 1994 (Survey #328).

Macallair, Dan, "Lock 'Em Up Legislation Means Prisons Gain Clout," *Christian Science Monitor*, September 20, 1994, p. 19.

Marcus, Ruth, and Ann Devroy, "Police Group Gives Bush Its Blessing," *The Washington Post*, October 10, 1992, pp. A1, A10.

Marion, Nancy E., *A History of Federal Crime Control Initiatives, 1960–1993* (Westport, Conn.: Praeger, 1994).

Martinson, Robert, "What Works?—Questions and Answers About Prison Reform," *The Public Interest* 35 (1974), pp. 22–54.

Masci, David, "$30 Billion Anti-Crime Bill Heads to Clinton's Desk," *Congressional Quarterly*, August 27, 1994, pp. 2488–2493.

Matusow, Allen J., *The Unraveling of America: A History of Liberalism in the 1960's* (New York: Harper Torchbooks, 1984).

Mauer, Marc, "The Fragility of Criminal Justice Reform," *Social Justice* 21, 3 (Fall 1994), pp. 14–29.

———, "The Drug War's Unequal Justice," *The Drug Policy Letter*, (Winter 1996), pp. 1–6.

Mauer, Marc, and Tracy Huling, "Young Black Americans and the Criminal Justice System: Five Years Later." The Sentencing Project, October 1995.

Mayer, William G., *The Changing American Mind: How and Why American Public Opinion Changed Between 1960 and 1988* (Ann Arbor: The University of Michigan Press, 1992).

McAdam, Doug, *Political Process and the Development of Black Insurgency, 1930–70* (Chicago: University of Chicago Press, 1982).

McAnany, Patrick D., "Assets Forfeiture as Drug Control Strategy: An Assessment." Paper presented at the Annual Meeting of the American Society of Criminology, New Orleans, Louisiana, 1992.

McCombs, Maxwell, and Donald L. Shaw, "The Agenda-Setting Function of the Mass Media," *Public Opinion Quarterly*, 36 (1972), pp. 176–187.

McCorkle, Richard C., "Punish and Rehabilitate?: Public Attitudes Toward Six Common Crimes," *Crime and Delinquency* 39, 2 (April 1993), pp. 240–252.

McGarrell, Edmund F., "Institutional Theory and the Stability of a Conflict Model of the Incarceration Rate," *Justice Quarterly* 10, 1 (March 1993), pp. 7–28.

McMillan, Edward J., "Diversified Resources Spell Success at ACA," *Corrections Today* 53 (June 1991), p. 70.

Meierhoefer, Barbara Stone, "The General Effect of Mandatory Minimum Prison Terms: A Longitudinal Study of Federal Sentences Imposed" (Washington, D.C.: The Federal Judicial Center, 1992).

Melossi, Dario, "Gazette of Morality and Social Whip: Punishment, Hegemony and the Case of the USA, 1970–92," *Social and Legal Studies* 2 (1993), pp. 259–279.

Merriam, John E., "National Media Coverage of Drug Issues, 1983–1987." In *Communication Campaigns About Drugs: Government, Media and the Public*, edited by Pamela Shoemaker (Hillsdale, New Jersey: Lawrence Erlbaum Associates, Publishers, 1989).

Michelowski, Raymond, "The Contradictions of Corrections." In *Making Law: The State, the Law and Structural Contradictions*, edited by William J. Chambliss and Marjorie S. Zatz (Bloomington: Indiana University Press, 1993).

———, "Some Thoughts Regarding the Impact of Clinton's Election on Crime and Justice Policy," *The Criminologist*, 18, 3 (May/June 1993).

Milakovich, Michael, and Kurt Weis, "Politics and Measures of Success in the War on Crime," *Crime and Delinquency* 21, 1 (January 1975), pp. 1–11.

Moore, David W., "Public Wants Crime Bill," *The Gallup Poll Monthly*, August 1994, p. 11.

Moore, Jonathan, *The Campaign For President: 1980 in Retrospect* (Cambridge, Mass.: Harvard Institute for Politics, 1981).

Morganthau, Tom, and Mark Miller, "Turf Wars in the Federal Bureacracy," *Newsweek* 113, 15 (1989), pp. 24–26.

Morley, Jefferson, "Crack in Black and White: Politics, Profits and Punishment in America's Drug Economy," *The Washington Post*, November 19, 1995, p. C1.

Morris, Lydia, *Dangerous Classes: The Underclass and Social Citizenship* (New York: Routledge, 1994).

Morris, Norval, *The Future of Imprisonment* (Chicago: University of Chicago Press, 1974).

Moynihan, Daniel Patrick, *The Politics of a Guaranteed Income: the Nixon Administration and the Family Assistance Plan* (New York: Random House, 1973).

Musto, David, *The American Disease: Origins of Narcotic Control* (New York: Oxford University Press, 1973).

National Institute on Drug Abuse, "National Household Survey on Drug Abuse: Main Findings" (Rockville, Md.: National Institute on Drug Abuse, 1988).

Nelson, Jack, "Bush Tells of Plan To Combat Drugs," *New York Times*, September 6, 1989, p. A11.

Neuman, Russell, Marion R. Just, and Ann N. Crigler, *Common Knowledge: News and the Construction of Political Meaning* (Chicago: University of Chicago Press, 1992).

New York Times/CBS News Polls. 1990. August Survey. New York: The *New York Times* Poll.

Niemi, Richard G., John Mueller, and Tom W. Smith, *Trends in Public Opinion: A Compendium of Survey Data* (New York: Greenwood Press, 1989).

Nimmo, Dan, *Newsgathering in Washington: A Study in Political Communication* (New York: Atherton Press, 1964).

Nixon, Richard, "If Mob Rule Takes Hold in the US: A Warning from Richard Nixon," *U.S. News and World Report*, August 15, 1966, pp. 64–65.

O'Brien, Robert M., *Crime and Victimization Data* (Beverly Hills, Ca.: Sage Publications, 1985).

Oliver, Charles, "In the Government's Ongoing War Against Crime, Money Has Been Little Object," *Investor's Business Daily*, May 12, 1994, p. 1.

Omi, Michael Allen, *We Shall Overturn: Race and the Contemporary American Right* (Dissertation: UC Santa Cruz, 1987).

Omi, Michael, and Howard Winant, *Racial Formation in the United States* (New York: Routledge and Kegan Paul, 1986).

Orcutt, James D., and Rebecca Faison, "Sex Role Attitude Change and Reporting of Rape Victimization, 1978–1985." In *Criminal Behavior: Text and Readings*, edited by Delos H. Kelly (New York: St. Martin's Press, 1994).

Orcutt, James D., and J. Blake Turner, "Shocking Numbers and Graphic Accounts: Quantified Images of Drug Problems in Print Media," *Social Problems* 40, 2 (May 1993), pp. 190–206.

Orfield, Gary, "Race and the Liberal Agenda: The Loss of the Integrationist Dream, 1965–1974." In *The Politics of Social Policy*, edited by Margaret Weir, Ann Shola Orloff, and Theda Skocpol (Princeton, N.J.: Princeton University Press, 1988).

Pear, Robert, "Reagan's Advisor's Giving Top Priority to Street Crime and Victim's Aid," *New York Times*, November 15, 1980, p. A1.

———, "FBI Chief Foresees Little Change Under Reagan," *New York Times*, November 23, 1980, p. A23.

Phillips, Kevin, *The Emerging Republican Majority* (New York: Arlington House, 1969).

Piven, Francis Fox, and Richard Cloward, *Poor People's Movements: Why They Succeed, How They Fail* (New York: Vintage Books, 1979).

Platt, Anthony M., "The Politics of Law and Order," *Social Justice* 21, 3 (Fall 1994), pp. 3–13.

Platt, Tony, and Paul Takugi, "Law and Order in the 1980s," *Crime and Social Justice*, Summer 1981, pp. 1–6.

Pomper, Gerald, "Toward A More Responsive Two-Party System," *Journal of Politics* 33 (1972), pp. 916–940.

Poveda, Tony G., "Clinton, Crime and the Justice Department," *Social Justice* 21, 3 (Fall 1994), pp. 73–84.

"President Forms Panel to Study Crime Problems," *New York Times*, July 27, 1965.

President's Commission on Law Enforcement and the Administration of Justice, *The Challenge of Crime in a Free Society* (New York: Dutton Books, 1968).

"Prison Privatization Is Efficient and Accountable, Wackenhut Corrections Chairman Maintains," *Financial News*, July 27, 1995, p. A12.

Quinson, Tim, and Robert B. Cox, "Making Crime Pay," *St. Louis Post-Dispatch*, February 10, 1994, p. C5.

Rasmussen, David W., and Bruce L. Benson, *The Economic Anatomy of a Drug War: Criminal Justice in the Commons* (New York: Rowman and Littlefield Publishers, Inc., 1994).

"Reagan: Drugs Are the No. 1 Problem," *Newsweek*, August 11, 1986, p. 18.

Reagan, Ronald, "Radio Address to the Nation on Proposed Crime Legislation,"

Public Papers of the Presidents 1984, volume 2 (Washington, D.C.: Government Printing Office, 1985), pp. 225–226.

———, "Remarks at the Annual Conference of the National Sheriffs' Association in Hartford, Conn.," *Public Papers of the Presidents of the United States 1984*, volume 2 (Washington, D.C.: U.S. Government Printing Office, 1985), pp. 884–888.

———, "Remarks at the Annual Convention of the Texas State Bar Association in San Antonio," *Public Papers of the Presidents of the United States 1984*, volume 2 (Washington, D.C.: U.S. Government Printing Office, 1984), pp. 1011–1016.

———, "Remarks at the Conservative Political Action Conference Dinner," *Public Papers of the Presidents 1983*, Volume I (Washington, D.C.: U.S. Government Printing Office, 1984), pp. 251–4.

———, "Remarks at a Fundraising Dinner Honoring Former Representative John M. Ashbrook in Ashland, Ohio," *Public Papers of the Presidents 1983*, volume 1 (Washington, D.C.: U.S. Government Printing Office, 1984), pp. 367–369.

———, "Remarks to Members of the National Fraternal Congress of America," *Public Papers of the Presidents 1986*, volume 2 (Washington, D.C.: Government Printing Office, 1987), pp. 1262–1264.

———, "Remarks to Members of the National Governors Association," *Public Papers of the Presidents 1988*, volume 1 (Washington, D.C.: U.S. Government Printing Office, 1989), pp. 235–239.

———, "Remarks at a White House Briefing for Service Organization Representatives on Drug Abuse," *Public Papers of the Presidents 1986*, volume 2 (Washington, D.C.: Government Printing Office, 1987), pp. 1022–1024.

———, "Remarks at a White House Ceremony Observing Crime Victims Week," *Public Papers of the Presidents 1983*, volume 1 (Washington, D.C.: U.S. Government Printing Office, 1984), pp. 552–555.

———, "Remarks at a White House Kickoff Ceremony for National Drug Abuse Education and Prevention Week," *Public Papers of the Presidents 1986*, volume 2 (Washington, D.C.: Government Printing Office, 1987), pp. 1329–1331.

Reaves, Brian A., "Drug Enforcement by Police and Sheriff's Departments, 1990: A LEMAS Report" (Washington, D.C.: U.S. Department of Justice, 1992).

Reeves, Jimmie L., and Richard Campbell, *Cracked Coverage: Television News, the Anti-Cocaine Crusade and the Reagan Legacy* (Durham: Duke University Press, 1994).

Reider, Jonathan, "The Rise of the Silent Majority." In *The Rise and Fall of the New Deal Order, 1930–1980*, edited by Steve Fraser and Gary Gerstle (Princeton, N.J.: Princeton University Press, 1989).

Reinarman, Craig, "The Social Construction of an Alcohol Problem: The Case of Mothers Against Drunk Drivers and Social Control in the 1980's," *Theory and Society* 17 (1988), pp. 91–120.

———, "The Social Construction of Drug Scares." In *Constructions of Deviance: Social Power, Context and Interaction*, edited by Patricia A. Adler and Peter Adler (Belmont, Ca.: Wadsworth Publishing Co., 1994).

Reinarman, Craig, and Harry Levine, "Crack in Context: Politics and Media in the Making of a Drug Scare," *Contemporary Drug Problems* 16, 4 (1989), pp. 535–577.

Reitz, Kevin R., "The Federal Role in Sentencing Law and Policy," *The Annals of the American Academy of Political and Social Science* 543 (January 1996), pp. 116–129.

The Republican National Party, "Republican Party Platform of 1968," *CQ Almanac* (Washington, D.C., 1968).

———, "The Offical Report of the Proceedings of the Republican National Convention" (Washington, D.C.: Republican National Committee, 1980).

Roberts, Julian V., "Public Opinion, Crime and Criminal Justice." In *Crime and Justice: A Review of Research*, volume 16, edited by Michael Tonry (Chicago: University of Chicago Press, 1992).

Roberts, Julian V., and Anthony Doob, "News Media Influence of Public Views on Sentencing," *Law and Human Behavior* 14, 5 (1990), pp. 451–468.

Roberts, Julian V., and Don Edwards, "Contextual Effects in Judgements of Crimes, Criminals and the Purpose of Sentencing," *Journal of Applied Social Psychology* 19, 11 (1989), pp. 902–917.

Rosch, Joel, "Crime as an Issue in American Politics." In *The Politics of Crime and Criminal Justice*, edited by Erika S. Fairchild and Vincent J. Webb (Beverly Hills, CA.: Sage Publications, 1985).

Rothman, David J., *The Discovery of the Asylum: Social Order and Disorder in the New Republic* (Boston: Little Brown and Co., 1990).

Ryan, Charlotte, *Prime Time Activism* (Boston: South End Press, 1991).

Sandys, Marla, and Edmund F. McGarrell, "Attitudes Toward Capital Punishment: Preference For the Penalty or Mere Acceptance?" *Journal of Research in Crime and Delinquency* 32, 2 (May 1995), pp. 191–213.

Sasson, Theodore, *Crime Talk: How Citizens Construct a Social Problem* (New York: Aldine de Gruyter, 1995).

Savelsberg, Joachim J., "Knowledge, Domination, and Criminal Punishment," *American Journal of Sociology* 99, 4 (January 1994), pp. 911–943.

Scammon, Richard, and Ben Wattenberg, *The Real Majority* (New York: Coward-McCann, Inc., 1970).

Scheingold, Stuart, *The Politics of Street Crime: Criminal Process and Cultural Obsession* (Philadelphia: Temple University, 1991).

Schiraldi, Vincent, "The Undue Influence of California's Prison Guards' Union: California's Correctional-Industrial Complex," San Francisco: Center on Juvenile and Criminal Justice, October 1994.

Schmitt, Eric, "Police Scorn Plan to Deny Schooling to Illegal Aliens," *New York Times*, April 9, 1996, p. A8.

Schram, Sanford, *Words of Welfare: The Poverty of Social Science and the Social Science of Poverty* (Minneapolis: University of Minnesota Press, 1995).

Schudson, Michael, *Discovering the News: A Social History of American Newspapers* (New York: Basic Books, 1978).

Schuman, Howard, Charlotte Steeh, and Lawrence Bobo, *Racial Attitudes in America: Trends and Interpretations* (Cambridge, Mass.: Harvard University Press, 1988).

Sears, David O., "Symbolic Racism." In *Eliminating Racism: Profiles in Controversy*, edited by Phylis Katz and Dalmas Taylor (New York: Plenum Press, 1988).

Secret, Philip E., and James B. Johnson, "Racial Differences in Attitudes Toward Crime Control," *Journal of Criminal Justice* 17 (1989), pp. 361–375.

Sellers, Martin P., *The History and Politics of Private Prisons* (London: Associated University Presses, 1993).

The Sentencing Project, *Facts About Prisons and Prisoners* (Washington, D.C.: The Sentencing Project, 1991).

Shapiro, Bruce, "How the War on Crime Imprisons America," *The Nation*, April 22, 1996, pp. 14–20.

Sigal, Leon V., *Reporters and Officials: The Organization and Politics of Newsmaking* (London: D.C. Heath and Co., 1973).

Simon, Jonathan, *Poor Discipline: Parole and the Social Control of the Underclass, 1890–1990* (Chicago: University of Chicago Press, 1994).

Smith, Anna Marie, *New Right Discourses on Race and Sexuality: Britain, 1968–1990* (Cambridge, Mass.: Cambridge University Press, 1994).

Speakes, Larry, "Press Briefing by Larry Speakes," The White House, Office of the Press Secretary, July 30, 1986, pp. 10–12. Simi Valley, Ca.: Document number 1847–07/30, Ronald Reagan Library.

Stinchcombe, Arthur T., Rebecca Adams, Carol A. Heimer, Kim Lane Scheppele, Tom W. Smith, and D. Garth Taylor, *Crime and Punishment in America: Changing Attitudes in America* (San Francisco: Jossey-Bass Publishers, 1980).

Stockman, David, *The Triumph of Politics* (New York: Avon Books, 1986).

Stolz, Barbara Ann, "Congress and Criminal Justice Policy Making: The Impact of Interest Groups and Symbolic Politics," *Journal of Criminal Justice* 13 (1985), pp. 307–319.

Stutman, Robert, *Dead on Delivery: Inside the Drug Wars, Straight From the Street* (Boston: Little Brown, 1992).

Sundquist, James, *The Dynamics of the Party System* (Washington, D.C.: The Brookings Institute, 1973).

Surette, Ray, *Media, Crime and Criminal Justice: Images and Realities* (Pacific Grove, Ca.: Brooks/Cole Publishing Co., 1992).

Taylor, D. Garth, Kim Lane Scheppele, and Arthur T. Stinchcombe, "Salience of Crime and Support for Harsher Criminal Sanctions," *Social Problems* 26, 4 (April 1979), pp. 413–424.

Thomas, Paulette, "Making Crime Pay," *The Wall Street Journal*, May 12, 1994, pp. A1, A8.

Thornton, Jeannye, and David Whitman, "Whites' Myths About Blacks," *U.S. News and World Report* 113 (November 9, 1992), pp. 41–44.

Tonry, Michael, "Racial Politics, Racial Disparities, and the War on Crime," *Crime and Delinquency* 40, 4 (October 1994), pp. 475–494.

———, *Malign Neglect: Race, Crime and Punishment* (New York: Oxford University Press, 1995).

Treaster, Joseph, "Emergency Room Cocaine Cases Rise," *New York Times*, October 24, 1992, p. A6.

———, "Emergency Hospital Visits Rise Among Drug Abusers," *New York Times*, April 24, 1993, p. A8.

Tuchman, Gayle, *Making News: A Study in the Social Construction of Reality* (New York: Free Press, 1978).

U.S. Department of Justice, *Attorney General's Task Force on Violent Crime: Final Report* (Washington, D.C.: U.S. Government Printing Office, 1981).

———, *The Sourcebook of Criminal Justice Statistics* (Washington, D.C.: Bureau of Justice Statistics, 1988).

———, *The Sourcebook of Criminal Justice Statistics* (Washington, D.C.: Bureau of Justice Statistics, 1989).

———, *The Sourcebook of Criminal Justice Statistics* (Washington, D.C.: Bureau of Justice Statistics, 1991).

———, "Fact Sheet: Drug Data Summary" (Washington, D.C.: Bureau of Justice Statistics, 1991).

———, *The Sourcebook of Criminal Justice Statistics* (Washington, D.C.: Bureau of Justice Statistics, 1994).

U.S. Office of the National Drug Control Policy, *National Drug Control Strategy* (Washington, D.C.: Executive Office of the President, 1992).

U.S. Sentencing Commission, "Mandatory Minimum Penalties in the Federal Justice System: Special Report to Congress" (Washington, D.C.: Government Printing Office, 1991).

Von Hirsch, Andrew, *Doing Justice* (New York: Hill and Wang, 1976).

Walker, Samuel, *A Critical History of Police Reform: The Emergence of Professionalism* (Lexington, Mass.: Lexington Books, 1977).

Wasburn, Philo C., *Broadcast Propaganda: International Radio Broadcasting and the Construction of Political Reality* (London: Praeger, 1992).

Webb, Gary, "Did Cops Get a License to Steal?" *Sacramento Bee*, September 26, 1993, p. F1.

Webster, William, "The Fear of Crime: Renew Our Vision of Community," *Vital Speeches of the Day* 47, 9 (1981), pp. 283–285.

Weld, William F., "Handbook on the Anti–Drug Abuse Act of 1986" (Washington, D.C.: U.S. Government Printing Office, 1987).

Wellman, Andrew, *Portraits of White Racism* (Cambridge: Cambridge University Press, 1993).

Whitney, Charles D., Marilyn Fritzler, Steven Jones, Sharon Mazzarella, and Lana Rakow, "Source and Geographic Bias in Television News 1982–4," *Journal of Electronic Broadcasting and Electronic Media* 33, 2 (1989), pp. 159–174.

Wilson, James Q., *Thinking About Crime* (New York: Basic Books, 1975).

Winant, Howard, "Difference and Equality: Postmodern Racial Politics in the

U.S." In *Racism, the City, and the State,* edited by Michael Keith and Malcolm Cross (New York: Routledge, 1993).

Wisotsky, Steven, "Exposing the War on Cocaine: The Futility and Destructiveness of the Prohibition," *Wisconsin Law Review* (1983), pp. 1305–1426.

———, "Crackdown: the Emerging 'Drug Exception' to the Bill of Rights," *Hastings Law Journal* 38 (1987), pp. 889–926.

Woodiwiss, Michael, *Crime, Crusades and Corruption: Prohibitions in the US, 1900–1987* (London: Pinter Publishers, 1988).

Wright, Kevin, *The Great American Crime Myth* (Westport, Conn.: Greenwood Press, 1985).

Wright, Paul, "The Federal Crime Bill," *Z Magazine,* March 1995, pp. 36–40.

Zimring, Franklin, and Gordon Hawkins, "The New Mathematics of Imprisonment," *Crime and Delinquency* 34, 4 (October 1988), pp. 425–436.

———, *The Scale of Imprisonment* (Chicago: University of Chicago Press, 1991).

———, *The Search for a Rational Drug Control Policy* (Cambridge, Mass.: Cambridge University Press, 1992).

———, *Incapacitation: Penal Confinement and the Restraint of Crime* (New York: Oxford University Press, 1995).

Index